Strategies for integrated communications

Strategies for integrated communications

State of the Art Report 12:5

Published by Pergamon Infotech Limited,
Maidenhead, Berkshire, England.

Printed by A Wheaton & Company Limited,
Exeter, Devonshire, England.

UDC 681.3
Dewey 658.505
ISBN 0 08 028 5902

Contents

Foreword

Technology is a good servant but a bad master. This aphorism makes itself apparent when an installed system is found to be simultaneously both indispensable and inflexible.

Many computer-based systems are installed in order to automate what have previously been manual systems.

> *Burren:*
> *The development of the IC has had major effects in fields other than computing. The large drop in the cost of circuitry has moved and continues to move both electrical and mechanical devices away from the use of analogue processes towards the use of digital processes. Typical examples are printing devices, sound handling and video.*

All too often, though, this process has occurred in a piecemeal and uncoordinated fashion within an organisation. Why should it matter that the main 'data processing' operations are run by the EDP department but that the word processors are bought by the office managers? The answer to this question is that this lack of coordination tends to lock up vital company information in different and incompatible systems. As information is one of the essential commodities or company resources it ought to be possible to assess its monetary value. It may be difficult or even impossible to assess the intrinsic value of information, but its loss or unavailability represents lost revenue, and its replacement or duplication costs time and effort.

At the point when a procedure cannot be carried out according to the user's needs but has to be done the way the computer system dictates, it is often too difficult or too expensive to make the required changes to the system. The investment in existing systems is frequently so large that reorganisation becomes a major undertaking. This situation is exacerbated by the rapidity of technological innovation (eg the appearance of the personal computer) which often leads to users tackling the 'symptoms of the disease'. The purchase of a personal computer may allow a user to process information in the way he chooses, but usually results in further incompatibility and fragmentation overall. Considerable ingenuity and effort is wasted in coping with 'the system', much of this being done in 'fire fighting' mode rather than with a long-term strategy in mind.

The information an organisation uses is not static. It changes and it needs to flow. The importance or value of a piece of information may endure or it may be ephemeral. This information may be communicated locally or over long distances (telecommunication) and may be communicated in the following forms:

- Data
- Text
- Picture
- Voice.

Three pivotal activities can be perceived:

1 Information management.
2 Information processing.
3 Information communication.

Dewis:
Istel's information communication strategy is one of its IT strategies, but what exactly is 'information communication'? 'Information' is a global phrase used to encompass a wide range of traffic types including data (structured format, for example as output on a line printer in an EDP environment), text (unstructured, including telex, teletex and communicating word processor traffic), image (eg public and private videotex service, facsimile, etc) and speech.

The manufacturers of computer systems and information processing products have traditionally tackled the various aspects of this market in isolated areas, but they are increasingly aware of the opportunities for integration. However, the tendency has been for each to take a parochial or closed view with respect to the product line. Purchasers and users have suffered from the resulting lack of compatibility between systems obtained from different vendors but have often been unaware of the consequences until recent times.

There is now a widespread recognition of the need for compatibility and standards, but the initiative has tended to come from the suppliers rather than the users (take, for example, the representatives on the committees of the International Standards Organisation (ISO) and the International Consultative Committee for Telephony and Telegraphy (CCITT)). Organisations using automated processing of information need to formulate an organisation-wide strategy for integrating their processes. This should be an organisational or business-driven activity, not a technology or end user-driven activity.

Burren:
The unanimous opinion of those working in the field is that there is no end in sight to the development of the IC. It is likely that over the coming years most electrical and mechanical devices used in the office, in the factory, in transport and in the home will be controlled by a stored program computer and will use digital techniques.

Technology represents both constraints and opportunities and the end user provides the requirements. The strategy must, of course, be sensitive to these things but should take the broad view and not be driven, willing or unwilling, by tactical problems.

The immediate question, to which there is no simple answer, is what should the strategy be? It depends on the individual organisation and what it is trying to do. However, the very realisation that a strategy is necessary is half the battle.

R Scantlebury: Editor

Publisher's note

This Report is divided into three parts:

1 Invited Papers.
2 Analysis.
3 Bibliography.

The Invited Papers in this State of the Art Report examine various aspects of strategies for integrated communications. If a paper cites references they are given at the end of the Invited Papers section, numbered in the range 1-99 but prefixed with the first three letters of the Invited Paper author's name.

The Analysis has the following functions:

1 Assesses the major advances in strategies for integrated communications.
2 Provides a balanced analysis of the state of the art of strategies for integrated communications.

The Analysis is constructed by the editor of the Report to provide a balanced and comprehensive view of the latest developments in strategies for integrated communications. The editor's personal analysis of the subject is supplemented by quotations from the Invited Papers and current literature, written by leading authorities on the subject.

The following editorial conventions are used throughout the Analysis:

1 Material in Times Roman (this typeface) is written by the editor.

2 Material in Times Italic (*this typeface*) is contributed by the person or publication whose name precedes it. The contributor's name is set in Times Italic. Numbers in parentheses in the ranges 001-099 or 100-199 following the name refer to the original source as specified in the Analysis references or the Bibliography, respectively, which both follow the Analysis. References within the text are numbered in the same way. A contributor's name without a reference refers to an Invited Paper published in this Report.

3 The quotations in the Analysis are arranged at the discretion of the editor to bring out key issues. Three or four dots within a single quotation indicate that a portion of the original text has been removed by the editor to improve clarity.

The Bibliography is a specially selected compilation of the most important published material on the subject of strategies for integrated communications. Each key item in the literature is reviewed and annotated to assist in selecting the required information.

Invited Papers

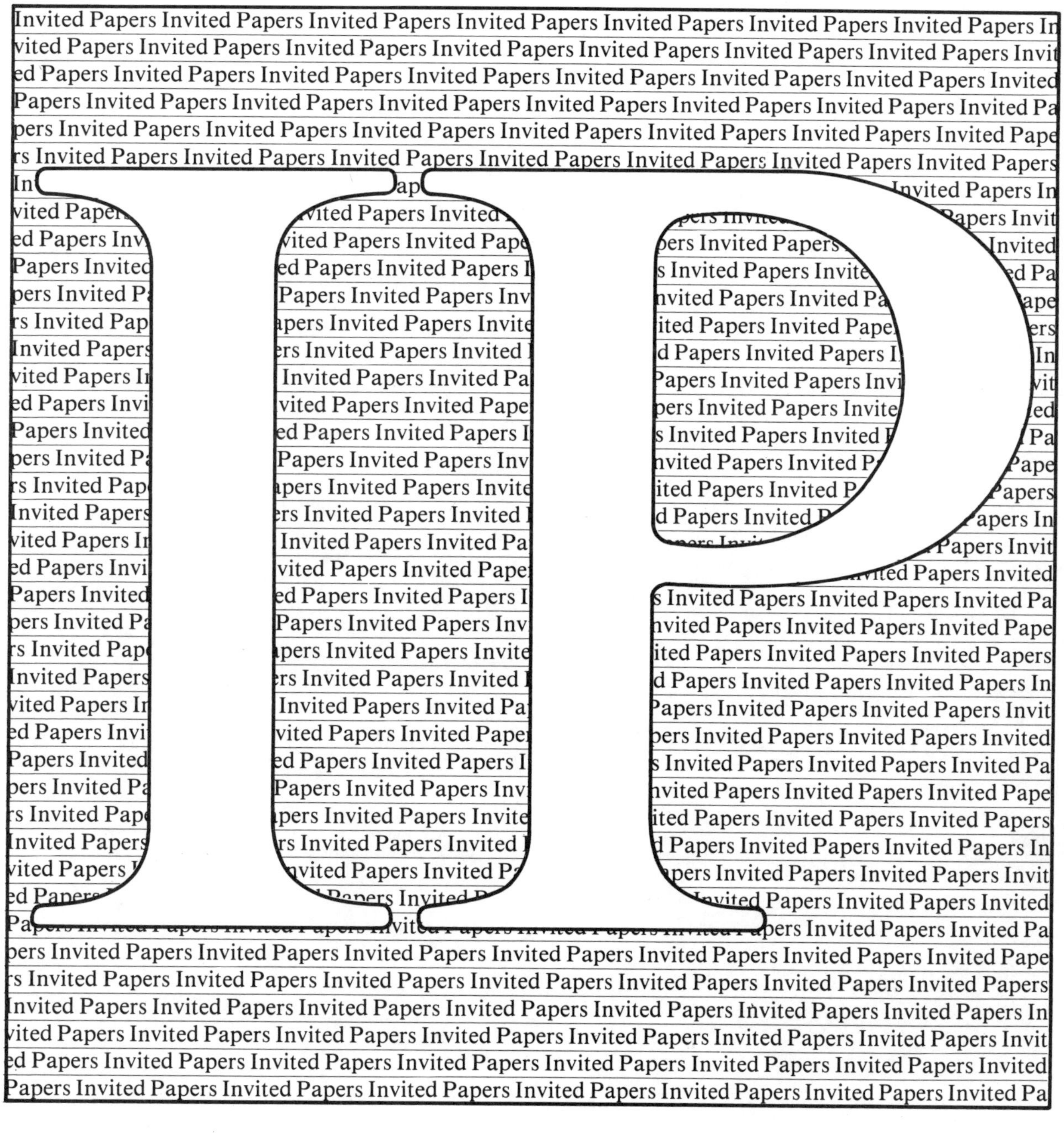

1: The interaction of computing and communications

J W Burren

Rutherford Appleton Laboratories
Didcot
Oxon
UK

The development of the integrated circuit has ensured an inevitable move away from analogue and towards digital systems of communication. This paper looks at the different types of traffic and the information technology systems that will be developed to deal with them. A description of the OSI model is included, as are details of the Project Universe experiment in high-speed computer networking for information technology.

J W Burren
Mr Burren graduated from Cambridge University in 1956 with a degree in Pure and Applied Mathematics. He worked in the theoretical physics division of AERE Harwell on the design of particle accelerators and, subsequently, at the Rutherford Laboratory on the problems of handling and analysing the vast volumes of data produced in high energy physics experiments. In the early 1970s he took charge of opening the Laboratory's large IBM computer centre to remote access from all UK universities. This work led to the development of SERCnet, the Science Research Council's X.25 packet-switched network. Since 1980 he has worked on Project Universe, a collaborative experiment in high-speed, wide-area networking.

The interaction of computing and communications

Introduction

Digital computer-to-computer communications traffic is currently carried, outside the confines of a single premises, by the telephone network. This network is designed to carry analogue voice signals and effectively limits transmission speeds to about 10 Kbit/sec with a fairly high bit-error rate. If challenged on this state of affairs, communications engineers will remind their computing colleagues that, in spite of all the fuss and excitement generated by computer networking, it is the telephone that is still by far the dominant source of telecommunications traffic. Thus the question 'Will the products of Information Technology, the combination of computing and communications, alter this situation?' There is an easy answer: in time the telephone will become a digital device and will itself be computerised. However, this paper will start by looking at other types of traffic and the information technology systems that will be developed to deal with them. Figure 1 lists some sources of traffic and gives equivalent volumes in terms of digitised voice at 64 Kbit/sec. For example, one A4 page of text is equivalent to half a second of voice and 'conferencing' quality colour TV is equivalent to about 30 voice channels.

It is interesting to look at a typical computer-based system that might generate substantial traffic, eg a facsimile transfer system working at 'xeroxing' quality. In this system the document being copied would be imaged onto a digitising system rather than directly onto the copier drum. The digital image would be transmitted over the network and then used to drive a scanning laser which would write the image onto the copying drum. From Figure 1 it will be seen that in compressed form an A4 page at 240 dots/inch, ie xerox quality, is equivalent to eight seconds of speech.

The computing structure required to support such a service is quite complex. The copying devices themselves would be computer controlled and would probably perform the compression and decompression of the data. One would in some circumstances operate the system as a real-time 'conversation' copier to copier. This mode would be useful in, for example, a teleconference situation, but in general it would be a great nuisance to have to synchronise the use of two copiers at different locations. Thus control computers with storage are required to organise the flow of data. In general, there would be an organising agent associated with each copier and these agents would be responsible for notifying the recipients of the arrival of facsimile information. The sending agent would also take responsibility for sending the document to multiple destinations. Further computers might be responsible for security services, such as encryption and authentication.

One could imagine that if such a service were to be made available on a wide scale, it would generate a significant volume of telecommunications traffic. Of course, the service would have to be available well ahead of the 'paperless' office!

The computing industry

Progress in computing has been dominated over the last decade and a half by the development of the

	bits or bit/sec	equivalent voice	compression factor (approx)	equivalent voice after compression
A4 page of text	32 Kbits	1/2 sec		
Voice	64 Kbit/sec	1 channel	5-10	
A4 facsimile black/white	5 Mbits	80 sec	10-15	8 sec
Graphics typical CAD	2 Mbits	30 sec		
Single-frame colour TV	4 Mbits	60 sec	3-5	12 sec
Colour TV	150 Mbit/sec	2400 channels	10-100	20-50 channels

Figure 1: Traffic generated by various media

Integrated Circuit (IC). Progress has been continuous and rapid. To take as an example one important computing component, the Random Access Memory (RAM), IBM made their first introduction of IC memory in place of core memory in 1973 with their second-round 370 machines, the 370/168 and 370/158. These machines used 1 Kbit RAM chips. Since then we have progressed through the 4 Kbit chip and the 16 Kbit chip to the current standard 64 Kbit chip and are about to enter the era of the 256 Kbit chip. Similar performance improvements have been made in processor and other types of chip. These improvements in technology have been accompanied by price decreases of the same magnitude.

The development of the IC has had major effects in fields other than computing. The large drop in the cost of circuitry has moved and continues to move both electrical and mechanical devices away from the use of analogue processes towards the use of digital processes. Typical examples are printing devices, sound handling and video.

The unanimous opinion of those working in the field is that there is no end in sight to the development of the IC. It is likely that over the coming years most electrical and mechanical devices used in the office, in the factory, in transport and in the home will be controlled by a stored program computer and will use digital techniques. The telephone is one such device. Although it is not possible to be precise about timescales, one can qualify 'over the coming years'. The computing industry is very competitive. It has large companies, even one dominant company, and no doubt these companies would like to control the pace of innovation. However, a combination of the Japanese and new companies in North America and Europe will ensure that no such control is possible.

The communications industry

Technical progress

Technical progress in this field over the last decade has been spectacular. The development of the IC has ensured an inevitable move away from analogue to digital systems of communiciation. Technical progress is viewed under three headings as follows:

- Transmission
- Switching
- Services

Transmission

Transmission is the technology of getting the 'bits' from place A to place B. The development of two new transmission technologies, optical fibre transmission and satellite transmission, has opened the possibility of huge increases in the volume of traffic that can be carried over both short and long distances. Following the inventions of the laser and the light-emitting diode in the early 1960s, the idea of digital signal transmission by optical fibre was put forward in 1966. At that time the idea was impractical as the attenuation produced by optical fibres was of the order of 1000 dB/km. (Note: dB is a measure of ratios on a logarithmic scale; in this case, of the signal power transmitted divided by the signal power received: dB = 10 log PT/PR. Thus a 3 dB loss corresponds to receiving only one-half of the transmitted power.) By 1970 fibres with attenuations of 20 dB/km had been produced to make the first practical demonstrations. First generation systems are now in operation using fibres with attenuations of less than 5 dB/km giving links carrying up to 140 Mbit/sec between repeaters separated by up to 10 km. Second generation systems that have been demonstrated in laboratories in the UK, US and Japan use longer wavelength sources to drive monomode fibres with attenuations as low as 0.2 dB/km. These systems can carry 500 Mbit/sec over distances greater than 30 km. Again there is no end in sight to development and no scientific discoveries are required to achieve systems carrying Gbit/sec over distances greater than 100 km without repeaters. Figure 2 gives some indications of the attractive properties of optical fibres.

Satellite transmission makes an excellent complement to fibre-optic transmission. Once established in the synchronous orbit a communications satellite makes possible the installation of high bit-rate transmission links between any points situated within a very large area, irrespective of the terrain separating the points. Water, mountain and desert are no obstacle and the areas covered can be both intracontinental and intercontinental. Satellite transmission can also be used for transmission links to moving objects, particularly to ships.

Advantages
Low attentuation long spacing between repeaters
High capacity hundreds of M bit/sec
Security, very difficult to tap
Small size
Lightweight
Immunity to electromagnetic interference

Disadvantages
Difficult to handle
splice, terminate, etc

Figure 2: Properties of optical fibres

Satellites do have the major disadvantage that they give a long transmission delay of about a quarter of a second. Additionally, as the synchronous orbit is above the equator the angle of elevation in countries well away from the equator is low; for example, in the UK the elevation is 30°. This makes transmission from land mobiles more or less impossible.

Typically, current satellites have channel capacities of 300-500 Mbit/sec and satellites with 2000 Mbit/sec capacity are planned. Again, very large increases in capacity are matters of engineering rather than science.

Switching

The situation in switching technology is not so clear. Again the IC is the key element. Electronically controlled circuit-switching exchanges have been developed. A variety of private exchanges are on the market and, in the UK, public network electronic exchanges (System X exchanges) are about to be introduced into the British Telecom (BT) network. The exchanges are all aimed at switching synchronous circuits with bit rates up to 64 Kbit/sec, the PCM bit rate for digital voice.

However, a different type of switching called packet-switching, developed for computer communications, is now also well established in Europe and North America, but on a very much smaller scale than the telephone network. Packet switching uses the telephone network for transmission but the exchanges are active computing elements either made up of multiple microprocessors linked together by some bussing structure or minicomputers with special line-handling equipment. Recently, a new form of distributed packet switching has come into prominence with the invention and development of the Local Area Network (LAN). In these networks, addressed packets are broadcast on a common channel to which all the clients have access. The switching function is distributed in the sense that only the client to whom a packet is addressed reads in the contents of the packet. The shared channel may be a ring of cable as in the Cambridge ring or a coaxial cable bus as in the ETHERNET technology. These schemes typically run the channel at 10 Mbit/sec. Development is proceeding on designs for 100 Mbit/sec and faster rings.

Services

For very many years the telecommunications services offered to the general public have stayed stable as telephone and telex. In the last few years there has been a move towards the development of new services. Three in particular are teletex, a computerised replacement for telex, videotex and picture videotex (known as Prestel in the UK) and videoconference.

Organisation

Traditionally, telecommunications provision in Europe and in much of the rest of the world apart from North America, has been organised as a state monopoly integrated with the postal system. These administrations have been extremely conservative in technical matters, working with massive long-term plans for the slow introduction of new technologies. These policies have produced major successes, for example international subscriber trunk dialling. However, the pace of change in the telecommunications industry has not kept up with activity within the computer industry. Something has to give. In the UK a start has been made — many would say a controversial start. Telecommunications has been separated from the postal services. The communications part, BT, has been given a potential competitor, Mercury, and in a rather short time has moved towards becoming a normal commercial enterprise working to maximise its business opportunities. It remains to be seen whether similar changes will be adopted in the rest of Europe.

The future

We thus have the position where there will be an enormous increase in the number of computer-driven devices over the next few years. Users of these devices will wish to exchange information in the form of text, graphics, data, images and voice. The basic technologies for transporting very large volumes of information are currently available and have enormous growth potential. The corresponding switching technologies are not so well developed, but the essential element for electronic switching, the IC, is well developed and has huge potential for further development. How then are the communications demands of these huge numbers of computer-driven devices to be met?

The two major models whose designs have been established to meet this requirement will now be considered:

- The Integrated Services Digital Network (ISDN)
- The Open Systems Interconnection (OSI) model.

Experience with the Universe network, which is an experimental packet-switched ISDN, will then be described.

ISDN

For a number of years telephone companies throughout the world have been planning moves towards digital telephone networks. In a fully digital telephone network, voice would be digitised by the telephone handset and this signal would be transmitted digitally through the network with switching being carried out by digital exchanges. The rate chosen for digitised voice is 64 Kbit/sec and since this rate is sufficiently high to carry many types of non-telephone services, the idea of evolving the digital telephone network into the ISDN came about. The ISDN would carry a wide range of communications services including both voice and non-voice services. Current proposals for the ISDN are to provide a standard customer interface consisting of two 64 Kbit/sec duplex channels together with a 16 Kbit/sec duplex channel for signalling. The 64 Kbit/sec channels may be used in either a circuit-switched or a packet-switched fashion: the signalling channel is packet switched.

Given that there is insufficient knowledge to build networks on the scale of existing telephone networks using packet-switched trunk exchanges, the ISDN proposals seem sensible. One has the feeling that too heavy building in of 64 Kbit/sec would be a mistake. Computer people will see warning parallels with computer architectures having address ranges of 64 Kbytes! Provision of broadband channels is envisaged in the ISDN proposals and is currently under study. One hopes that recommendations will be agreed in the near future.

In the UK a pilot ISDN service is due to be opened in London in mid-1984. This is based on System X exchanges and will offer one 64 Kbit/sec channel for voice or data, plus one 8 Kbit/sec data channel, plus an 8 Kbit/sec signalling channel. The access service to the ISDN is to be called Integrated Digital Access (IDA). The service will be extended quite rapidly to Birmingham and Manchester. Further extensions to the service will follow the general pattern of the installation of System X exchanges.

OSI model

As is very well known, the International Standards Organisation (ISO) has been developing a standard reference model and standard protocols for the exchange of information between systems. The OSI model divides the functions required to organise the exchange of information between systems into a set of seven layers. The layers are arranged in a hierarchy with the bottom layer being concerned with questions of physical access to the transmission medium and the top layer being concerned with applications. Standards are being developed to fit into the various layers of the model.

The model allows standards activities on various aspects of systems interconnection to be pursued independently of one another and aims to allow protocols to support new technologies and new applications to be slotted into the appropriate layers of the model without disturbing the whole protocol structure. It is a very ambitious project; the intention is to cover the interconnection activities between systems involved in all types of application. The model has already had a very large influence on all aspects of computer communications. The field of application of the model is in a sense orthogonal to that of the ISDN and it is the intention of CCITT that the ISDN should conform to the OSI model.

Clearly, the idea of producing an orderly set of standards covering a very wide range of applications involving communications between computing systems is very sensible. For this reason the ISO model and the ISO standards have become a formidable driving force in the development of computer systems. There are, however, some difficulties with the model. Hierarchical structures have the general difficulty that they assign a fixed ordering of the functions and this ordering may be unsuitable to changes in application or technology. There have already been difficulties in accommodating LAN and satellite transmission technologies within the model.

A second difficulty is that the model and standards do not take account of speed. While networks are based on the analogue telephone network operating at speeds of 10 Kbit/sec this is not important. However, there are limits to the speed at which messages can be exchanged between systems, limits which are set ultimately by the velocity of light. Protocols that involve a heavy exchange of messages, and generalised protocols do tend to involve such exchanges, place limits on the rate of transfer of useful information. It remains to be seen whether the ISO model and protocol standards can accommodate applications requiring high rates of information transfer.

Finally, the model and the current set of standards under development do not cater for voice!

Project Universe

Project Universe is an experiment in high-speed computer networking for information technology. The Universe network, which has been built as part of the project, is a high-speed packet-switched network servicing experimenters at seven sites: three industrial establishments, BT Research Laboratories, Marconi Research Centre and Logica; three universities, Cambridge, Loughborough and University College London (UCL); and the Rutherford Appleton Laboratory. Apart from the three companies and the SERC, the Department of Trade and Industry has been a major source of funding.

Each of the seven sites is equipped with one or more LANs and long-distance connectivity is achieved primarily by use of a broadcast satellite channel using the Orbital Test Satellite (OTS). Six of the sites have earthstations and use the satellite channel; the seventh site, at Logica Ltd in London, is connected to UCL by an Mbit/sec cable link. There are 12 LANs incorporated in the Universe network, 11 of these are Cambridge rings and one is a token ring. Attached to these LANs are some 150-200 computers scattered over the seven sites.

The transmission media used in the Universe network, the LANs, the satellite channel and the terrestrial cable share the characteristics of high bit rate (Mbit/sec) and low bit-error rate (one in 10^9 or better). Both the LANs and the satellite channel provide switching by broadcast. Thus stations connected to the media transmit addressed packets which are broadcast to all stations connected to the media. On receipt, stations check the address and only the station or stations with the correct address read in the packet. Various schemes are used to prevent more than one station transmitting at the same time, which would corrupt the information. In the Universe network the various media are linked by computers, called bridges, which are simultaneously connected to two of the media (see Figure 3). There are three types of bridge; ring-to-ring, ring-to-satellite and ring-to-cable.

Project Universe is concerned with both the construction of the network and with carrying out a number of application experiments to assess the performance of the network and to determine the optimal strategies for using the network as the basis for a wide range of information technology applications. The applications cover not only traditional DP activities but new classes of data such as digitised voice and pictures.

The Universe network offers its clients the communications functions of a packet-switched ISDN. Three types of traffic are supported; datagram, exchange and connection. Datagrams are packets that are routed individually through the network independent of other traffic (apart from loading conditions). Packets in the network may be up to 2000 bytes in length and datagrams are intended primarily for monitoring activities. Exchanges consist of a single block request in one direction followed by one or more blocks of reply. Connections are duplex paths between clients down which many blocks may be sent in either direction. Connections are similar to circuit-switched connections providing sequentiality, ie preserving the order in which information is sent, but without the value-added services of retransmission after error and flow control usually associated with packet networks. In addition to these basic transmission facilities, an infrastructure for the management of distributed computing activities across the network has been built up within the project. This infrastructure consists of substantial numbers of both dedicated and general-purpose 'server' computers built out of standard microprocessors. For example, all services on the Universe network are known by name and tables for converting names to locations (addresses) are held in dedicated servers called 'nameservers', which appear on each ring.

The Universe network was built and commissioned during 1982. During 1983 it has been used to carry out a series of applications experiments together with a program of performance measurement.

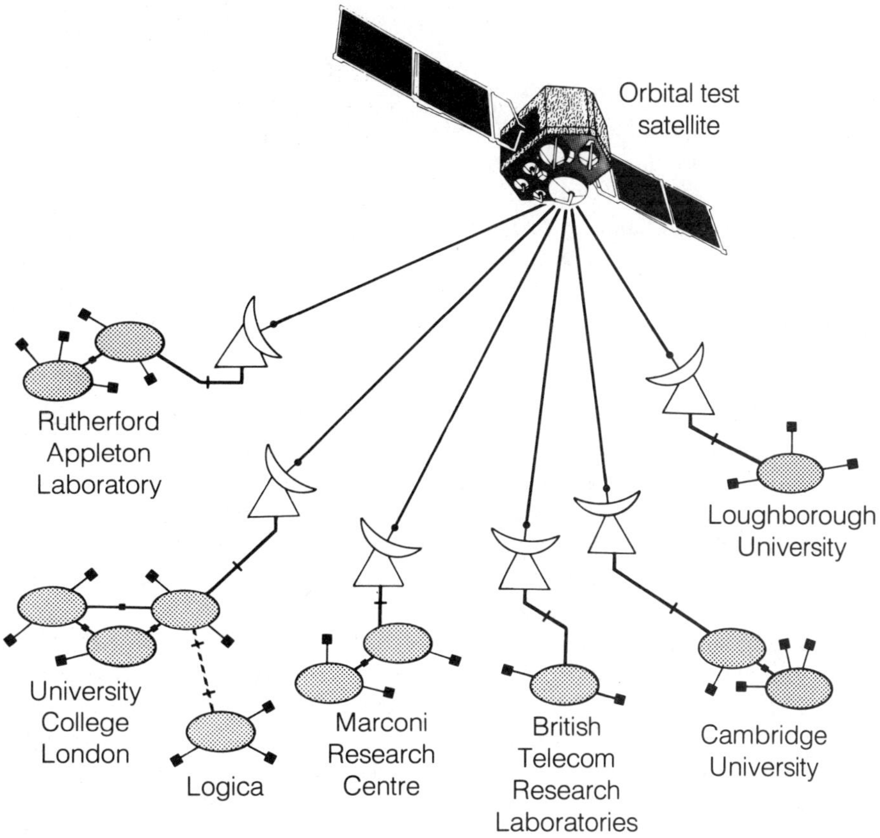

Figure 3: The Universe network

Experiments have included the following:

1 The transmission of digitised voice both for real-time conversations and for a stored voice-message system.

2 The transmission of static colour video images and slow-scan moving images. (The network does not have a high enough bit-rate for full moving video.)

3 Management of the flow of information between clients, including facilities for encryption and authentication and for managing the situation in which the destination is only intermittently available, as will be the case when personal workstations are connected to an office network.

4 Work is also in progress to study the way in which software engineering tools need to be upgraded when the development environment includes resources at a number of different locations. The aim here is to provide a uniform working environment to the developer who is using a wide range of distributed facilities.

In addition to the experimental program, the project has mounted three public demonstrations of the Universe network during 1983. Two of these, INFO'83 and Business Telecoms, have been at the Barbican Centre and the third at World Telecoms in Geneva. Each demonstration has involved setting up a complete Universe site including a large ring with upwards of 15 computers attached together with a satellite bridge and mobile earthstation to give connectivity to the other sites. To reach the earthstation site, the ring at the Barbican had to cover two locations about 1 km apart, giving a ring some 2 km in length.

Transmission over the long sections of the ring was by means of fibre-optic cables. At Geneva the exhibit was divided between two locations about 200 metres apart and again fibre-optic transmission was used for the long sections. These exhibition sites have to be installed and operational within a few days and demonstrate the maturity of the technology. A full range of applications experiments were shown at each exhibit, including packet voice, colour video, image transmission and text, graphics and image retrieval.

Commercial exploitation

Successful commercial exploitation of the Universe-type network depends both on having a viable network technology and on having commercially interesting applications that exploit the capabilities of the network. In a second phase of the Universe Project it is hoped to move the network to an experimental service stage by introducing a number of megastream links between the sites. Megastream is a 2 Mbit/sec digital point-to-point service recently introduced by BT. At the same time the individual applications that have been developed in the first phase of the project will be integrated into 'quasi-service' systems for the integrated office environment, for distributed software development and for teleconferencing.

Conclusion

For those involved in the technology of computing or communications, we live in very exciting times. In both computing and communications the discoveries and innovations come thick and fast and this state of affairs seems set to continue in the foreseeable future. There are forces that would like to control the pace of change but these forces are certainly not dominant in the computing industry. In the communications industry organisational change is also in the air. For many, changes in this area are most unwelcome and it is not clear whether or not the forces against change will, in the end, dominate.

However, there is a strong view that it does not make technical or economic sense to stop communications providers from providing computing services and equipment. In particular, this view seems to have won the day in the US. If this view is held then an almost inevitable consequence is that communications provision must move towards an organisational structure similar to that of the computing industry, ie away from state administration towards competitive business. There are formidable problems to be solved (no one, for example, wants to see their telephone supplier go bankrupt) and these are made harder by the political overtones. However it is surely not beyond our ingenuity to devise satisfactory solutions in the coming years.

2: Regional, urban, national and international communication systems

B C Burrows

Future Information Associates
Milton Keynes
UK

Voice and data can be transmitted over the same network and, by the use of digital technology, facsimile is now being integrated into the system. Graphics could also be transmitted via the same network in the near future. A conflict is developing where cable is used with the concept of the wired city and the wired nation, and when microwaves, satellites and earthstations are used. This paper looks at how these systems develop, conflict and integrate, on a regional, national and international basis.

B C Burrows

Mr Burrows is a consultant in the areas of corporate planning, new production promotion, new technology development, information technology and the property market. He is a member of the Society for Strategy and Long Range Planning, Aslib, the Institute of Information Scientists and the Industrial Marketing Research Association. He has written various publications on the subject of urban planning information research services, information technology and the property market ('Long-range planning', August 1984).

Regional, urban, national and international communication systems

Introduction

There are two aspects in the technical development in communication networks. Firstly, new methods such as the use of fibre optics, area networks and teleconferencing are developing at an increasing pace and thus creating new services. At the same time, these services are converging into a common technology by the use of digital transmission. For example, both voice and data can be transmitted over the same network and, by the use of digital technology, facsimile is being integrated into this system. This means that graphics could soon be integrated into the same system as voice and data communications.

A conflict is developing when cable is used with the concept of the wired city and wired nation and when microwaves, satellites and earthstations are used. An attempt will be made to see how these systems develop, conflict and integrate, on a regional, national and international basis.

The UK

The UK, in common with most developed countries, used an analogue telephone network for voice transmission, ideally suited for the purpose but not suitable for the transmission of data: the telegram is now obsolete and the standard telex machine is both slow and inefficient. Facsimile and electronic mail are being introduced as alternatives; it is now possible to use any shared logic word processor for electronic mail, with the likely possibility of integration with facsimile in the near future.

British Telecom (BT)

The telegraph preceded the telephone system by a long period. In fact, the Post Office was given a monopoly by the Telegraph Act of 1868 *(BUR1)*. The telephone system was originally developed by a private company, the National Telephone Company, and in 1911 it was transferred to the Post Office. This had advantages in that an overall service could be provided but because all new equipment had to be approved and tested by the Post Office, the implementation of the new technology was slow. As a result of much discussion on this, the Beesley Report was produced as well as the setting up of BT as a separate body from the postal services.

The Beesley Report

There were two main conclusions from this Report *(BUR2)*:

1 Other organisations should be able to compete with BT in the non-voice communication market.
2 BT would be able to enter and compete in this market.

This Report was accepted by the Government and it is interesting to examine its effect in the market both in relation to BT and to firms in the private sector.

The effect on BT

BT has an excellent research establishment at Martlesham, near Ipswich. However, the acceptance of new equipment from the private sector has either been delayed or rejected as a result of BT's monopoly. BT has also sometimes been slow to respond to the needs of the customer. Most BT workers now recognise this and welcome the new challenge of this liberalisation policy — in fact, its range of services has been greatly extended and improved. New services include a digital network, System X, which is operational in some areas; other areas have been linked to this with fibre optics. The ultimate plan is to have a national digital fibre-optic network. Other services include Satstream, which is a small dish satellite service, and teleconferencing services.

The most recent plan is to privatise BT, although there remains some controversy about this. Meanwhile, the Government has also given a licence to a rival system, Project Mercury.

Project Mercury

Project Mercury is a consortium set up by Cable and Wireless, Barclays Bank and British Petroleum*. The plan is to run fibre optics along the British Rail track in order to link the major cities with a fast communications system. A licence has been granted and BT has agreed to link this into its system. One contending problem is which system of communication is to be used when one reaches a terminal station. The solution is to use dish aerials and microwaves.

Wayleaves

A wayleave is a legal term meaning permission to have access in, over or under someone else's land. The power for this rests with the local authority. Examples are sewage and water services, telephone wires and electrical cable.

It is interesting that since liberalisation, new wayleaves have been found. We have seen the example of the use of railway lines; Mercury is looking at the possible use of motorway verges and towpaths on canals, and it also attempted and failed to get permission to use the underground inner circle railway in London.

Central Electricity Generating Board (CEGB)

There is a planned project by CEGB in conjunction with Plessey and BICC to run fibre optics along the overhead grid system *(BUR3)*. It is still at an experimental stage with trials at Southampton. One of the problems is of what happens when the power lines go underground near and in urban areas; dish aerials at the end of the grid system linked to satellites have been suggested.

City sewers

The latest proposal is to run fibre optics along the sewers. As these are glass there is no problem of deterioration. This project has been widely reported *(BUR4 — BUR7)*. A consortium has been set up by the Water Research Centre and a company called Cabletime. This company is a joint venture between Times Fibre Communications Inc in the US (holding company Insito Inc) and United Engineering Industry (UEI) Ltd in the UK. UEI is a holding company for a number of high technology companies which include Link Electronics, Micro Consultants and Quantel Ltd. There are plans for trials in Swindon. This seems an excellent idea and is of considerable interest to the planner engaged in urban renewal. The older cities have sewers which were put in many years ago and need renewal. Vast sums of money are needed to replace the sewers before starting a rebuilding programme. The rent paid to use these sewers could help to renew them and release money for rebuilding. It could also establish a first-class local area network which would attract industry and commerce, all badly needed in older cities.

---

*Barclays Bank and British Petroleum have since withdrawn.

Satellites and microwaves

Both BT and Mercury have earthstations operating in the UK. They are becoming cheaper to construct because satellites are becoming larger and, since less land is required, their siting is becoming less controversial. BT is also developing smaller, lighter and cheaper dish aerials. This means that it will be possible to place these on roofs of commercial buildings without any structural alterations.

Wide Area Networks (WANs) and Local Area Networks (LANs)

The dividing line btween WANs and LANs is not obvious. For example, the plans to link universities into a communications network via the Project Universe scheme comprise a WAN; if a university used ETHERNET to link its equipment together it would be using a LAN. It is likely that some central business districts will soon be wired up with fibre optics. Is this a LAN or a WAN? The author's view is that it is a LAN, but in order to encourage future development, a common terminology is needed. Many WANs using microwaves will be able to link into LANs using cable.

Alternative systems

The use of alternative systems is important for the business sector. The licensing of the two rival systems for a set period is a good idea enabling developments to be monitored. It is likely that most large organisations will use both systems so that if one fails, the other system can be used. This trend will be set by the firms engaged in the financial sector and could also result in the decline in use of the leased line.

Videotex systems

When videotex systems were launched, the Post Office was not able to register the name 'Viewdata' because it was too descriptive. Therefore, the name 'Prestel' was registered. Faced with other countries developing their own systems, the CCITT established the term 'videotex' as the word to be used for all systems. However, 'viewdata' is still used in the UK.

The concept of viewdata was devised by Sam Fedida *(BUR8)* when working for the Post Office. The basis of this invention is to link the telephone with the TV set so as to use this as a terminal for receiving information. This was slow to develop for the following reasons:

1 It was seen as a means of stimulating the use of domestic telephones which were underutilised.
2 Responsibility for collecting and providing the information was left to other organisations.

The result, as predicted by most members of the information profession, was that marketing targets were not met. BT, which now had responsibility for Prestel, realised this and targeted the business sector. Although Prestel cannot challenge the bibliographic and economic on-line information system, as it is too well established, it is developing Gateway, to enable on-line systems to be accessed via Prestel.

However, the main development of Prestel has been to create a new market not covered by on-line systems. This is mainly transitory information in the following areas:

- Hotel booking
- Travel — tourism
- Banking — electronic funds transfer
- Teleshopping
- House purchase — estate agents
- Advertising
- Education
- Community information.

Of these, the tourist industry seems to be the area of most growth. Many travel agents are now using Prestel.

Despite BT's claims, there is not much information on Prestel. What there is has a retailing and service bias. Closed user groups have been set up mainly because of lack of information within the system.

Closed user groups

Closed user groups are being set up in four areas:

1 Communications within an organisation. For example, Thorn EMI has put in a system at all its locations in the UK in order to improve communications.

2 Community information. This has developed in several areas; some have been developed by Citizen's Advice Bureaux, some via public libraries, some via cooperative schemes such as LASER; and some have been developed by a commercial organisation, often a local newspaper publisher.

3 Specialist groups. These are likely to develop via professional services, ie lawyers, solicitors, etc.

4 Extensions of business; for example British Leyland service to dealers.

Mailbox

Mailbox is a system in which messages can be transmitted to another user of Prestel. This is stored in the system so that it is available to the person concerned when the TV set is used. It is a form of electronic mail and could develop with word processing and teletex via videotex systems.

Because of the mistake in marketing, the UK has lost its lead and other countries have developed rival systems. However, the increasing use by the service sector of both Prestel and closed user groups should give the UK an advantage, with Germany the main rival in this sector.

Cable TV

Cable TV started in the US in 1948 as a means of overcoming poor reception and to bring TV to areas not yet covered. Development followed the same pattern in the UK, Swindon being an example. It was also one of the reasons for installing cable in Milton Keynes. The other reason was an attempt to avoid the large-scale use of aerials for aesthetic reasons. As there was evidence of further development of CATV in the US, the response of the British Government was to set up an Information Technology Advisory Panel (ITAP) to report on this *(BUR9)*. The panel recommended the cabling of Britain for multichannel TV in the entertainment areas. There was no mention of business use. Also, coaxial cable was recommended and a degree of urgency implied.

The Hunt Committee was set up in response to this and after requesting evidence the ITAP proposals were accepted *(BUR10)*. The Institute of Information Scientists also submitted evidence *(BUR11)* recommending the development of cable TV for business use as well as a consideration to use fibre optics.

Now that these proposals have been accepted the Government has granted a series of licences to operate multichannel TV in specific areas. The one of interest is Westminster, where it is likely that the business sector could take the lead.

Urban and regional development

The large urban areas were originally created by industry and commerce in the Victorian era. Today, there is a migration of population from larger towns to the regional centres. This is also happening in the US and Japan. Commerce and industry have followed this pattern. It is certain that large cities will survive as important financial centres for international exchange but much of manufacturing is moving to regional areas. The concept of the science parks, linked to a university, means that research, production and administration are also being established in regional areas. The preserve of the manufacturing firms will still be retained by a sales office in the central business district.

Physical communication patterns

The most important physical communication patterns for advanced technology firms are based on the location of airports and motorways. For example, the M4 corridor has developed from Heathrow into Wales. The same pattern could happen for the M1 when the M25 ring road around London is completed in 1986.

The impact of new communications technology

Industry will continue to be located near good physical communications because both raw materials and finished goods have to be moved. The question to be asked is what impact will the new technology have in the central business district because people can now work from home and, by using on-line databases, have access to as much information as in an office. There is the example of Rank Xerox with plans for part-time working from home; also, increasing numbers of consultants are using on-line systems, and one of the software companies, F International, has most people working from home.

However, it is likely that the financial and service sector will remain within the central business district because of the informal 'old boy network' information system. The electronic office with local area retrieval will be used for internal needs as well as package switching and satellites for external communications.

International networks

International networks have been slow to develop because of the lack of cooperation between PTTs, particularly in Europe. Communications to the US are better due to the services provided by TELENET and Tymshare. The position in Europe is now greatly improved with the network set up by EURONET and by the EEC. This has been so successful that the PTTs are now switching the calls between countries.

There is an area of conflict between satellite and microwaves and cable. There are plans to replace existing cables across the Atlantic with a fibre-optic cable. However, it is likely that the two systems will continue together on an international basis for some time.

Canada

Cable penetration in Canada far exceeds that in the US. Over 50% of Canadian households have a cable service, and cable passes over 70% of all households *(BUR12)*. There are more regulations for cable use and development than in the US. Future development could depend on the use of fibre optics but no plans for this have been traced.

Canada has the most advanced videotex system, called Teledon. It is an alpha-geometric system, instead of an alpha-mosaic system, which results in a greatly improved presentation, particularly with graphics display. As the videotex market is not well developed in the US Teledon could become the system used there *(BUR8)*.

Holland

Holland, and multilingual countries such as Belgium, uses cable TV to a great extent. In fact, in Holland seven out of 10 houses can receive TV and radio by underground cable.

The British plan favoured by the Government is to put coaxial cable into ducting so that it can be replaced by fibre optics. The Dutch plan is to bury reinforced cable; it is claimed that this has a life of 40 years. An article in the 'New Scientist' favours the Dutch plan *(BUR13)*.

The Dutch Government has just announced two experimental cabling projects. At Geldrop, near Eindhoven, the Dutch Post Office, Philips and the universities of Delft and Eindhoven are planning an interactive fibre-optic system. There were plans to start laying optical fibres to Dutch houses by 1990 but this date has been put back to 1995.

In Limburg the Dutch Government and the local authority are paying half of a £40 million project to provide 100 000 homes with an interactive link based on coaxial copper cable. The point raised by the article in the 'New Scientist' *(BUR13)* is that the experience of the use of coaxial cable for a considerable period will be a great advantage when switching to fibre optics.

As it is clear from the evidence that fibre optics will be competitive with coaxial cable by 1986, it would therefore seem that further development of copper cable would not be worthwhile. Therefore, if Holland is to make a contribution to the concept of the wired nation the plans at Geldrop are the ones of interest, but a shorter timescale for development will be needed.

France

New methods of communications are being introduced in the UK by the Government encouraging the private sector to both develop and pay for these. The other European country with a specific approach to communications is France. It has a tradition of a centralised approach and the Government is promoting and paying for an advanced programme based on fibre optics. The plan is to wire up 320 000 homes for cable TV at a cost of 116 million by 1985/1986. It is also hoping to launch a direct broadcasting TV satellite. One of the areas to be cabled is Paris, wiring 46 000 homes by 1985 rising to 400 000 by 1992 *(BUR14)*.

Germany

Germany is well advanced with videotex in the service sector but its plans for CATV are only just emerging. It has just launched a system with 19 channels to 1200 subscribers. Cable TV is thus not likely to develop as fast as in the UK and France.

Europe

A study of cable TV in Europe as a whole was made by the 'Economist' *(BUR15)*. A table shows that the countries with the largest amount of cabling are the smaller multilingual countries such as Belgium and Switzerland. However, much of this cabling was installed a long while ago and is non-interactive coaxial cabling. Some of this could be converted to an interactive system but it would be very expensive. It is likely that development will be fastest in France and the UK.

As already mentioned, there is a consensus view that by 1986 fibre optics will be competitive with coaxial cable. Therefore, whoever has the most comprehensive broadband fibre-optic network which can be used both for leisure and business will have the most viable system.

US

The US was the first country to use cable TV in order to bring this service to areas not covered by transmission. Since then, there has been much development in the following areas:

- Multichoice TV
- Pressure groups
- Religious groups
- News service
- Continuous 'pop' music service
- Local news and community projects
- Soft pornography market.

Attempts have been made to regulate this market and make it more responsive to the community. The whole communication network in the US is, however, in a state of flux due to the breaking up of the AT&T

monopoly. Many smaller companies have new ideas and are playing a much greater role in the market *(BUR16)*. The concept of information utility was popular for a while *(BUR17)*. There have also been plans to run CATV as a municipal enterprise but this has been met with limited success *(BUR18, BUR19)*. It is often the entrepreneurs such as Ted Turner with his news service who have been the most successful. Overall, however, the market position is not clear; as soon as one company announces a successful year, another runs into financial difficulties. The successful US companies are now interested in exporting their technology and services to the European and Japanese market.

It is likely that the breaking up of MaBell will follow the pattern of the situation in the UK after the Beesley Report. It should be noted that the US has a lead with satellite technology and sees this as a means of exporting direct TV programmes as well as establishing a major share of the business sector. It also has a lead in videoconferencing.

Japan

In 1970, the Japanese Computer Usage Development Institute outlined 'The plan for an information society. A national goal toward the year 2000' *(BUR20)*. The section on communications is outlined below.

Nationwide information network

The current telephone lines will be employed temporarily and gradually replaced by a new data communication network employing coaxial cable and microwave with a capacity of one million data communication circuits by 1985.

Nippon Telegraph and Telephone (NTT)

The body responsible for implementing this new network is NTT, a public monopoly similar to BT. This organisation has behaved in a similar manner to BT in restricting the use of new equipment and there are now plans to 'decontrol' it in a similar manner to the plans outlined for BT in the Beesley Report.

Use of cable

So far, this has been used in only two areas:

- Tama New Town — coaxial cable
- Higaski Ikoma New Town H1-OVIS Project — fibre optics.

There are also plans in hand for a large-scale use of CATV in Tokyo.

The use of satellite and microwave for communications in Japan has not been given much consideration.

Firms in Japan in the telecommunications industry

There are four firms which control most of the area of telecommunications business:

- Nippon Electric Company
- Fujitsu
- Hitachi
- Oki Electric Industry.

The most recent development in telecommunications is a planned information network system. This will be a national digital network using fibre optics with a target date by the 1990s *(BUR21)*. The plans to privatise NTT have also been finalised *(BUR22)*. It is therefore clear that Japan has the most comprehensive retrieval plan and is a major challenge to the US and Europe.

Third World

The problems facing the Third World are considerable and one of the great barriers to economic growth in the Third World is the lack of good communications. One solution is to use satellites and the Japanese are now developing a satellite industry for this purpose. They are also developing photovoltanic solar cells by the use of amphorous semiconductors; these can act as a stand-alone power source independent of a national electric grid system.

While there is a migration of population from the larger towns to regional areas in the developed countries, the opposite is happening in the Third World. By the use of improved communication systems, it is possible that this process could be halted so that both the undeveloped and developed world could enter the post industrial society at the same time.

The future

International

We have seen that a more effective global communications system could result in a less divided world. The question is what system or systems will be used. It is likely that satellites will be used for a considerable period, particularly in the Third World. Eventually, however, their use will probably decline in favour of cable which, in the long run, is more economical.

Satellites will continue to be used for monitoring our environment. They can and are being used for forecasting weather patterns including hazards such as hurricanes. They are also used for monitoring pollution and the heating up of the atmosphere due to rising population and industrial growth, which could cause the 'greenhouse effect'. Other important uses are for irrigation studies and mineral exploration.

National, regional and urban areas

The concept of the wired nation and the wired city is likely to develop and should be well established by the year 2000. Japan has the most advanced plans but it is possible that the race could be between Japan and the UK. Both are small, densely populated islands, which have a considerable advantage. The key is the setting of agreed standards, plans for which are in hand via Alvey and the consortium set up by ICL. Also, CCITT has established well recognised standards. These are vital for future development.

The UK

If the UK succeeds in becoming the first wired nation, it will have the following advantages:

1 It will have the most efficient communications network in the world, which will stimulate economic growth and investment.

2 There will be a major export market for this equipment.

It is worth looking at some of the projects in the UK in order to study areas of possible growth.

Mercury/Cabletime
As we have seen, Mercury currently uses microwaves for information within the city. One alternative would be to link up with Cabletime and use the sewers. An item in the 'Financial Times' *(BUR23)* consists in an interview with Peter Michael, representing the British end of Cabletime, in which it was stated that discussions are taking place with Mercury.

BT/Mercury
It is likely that by the year 2000 BT/Mercury could have a fibre-optic national network. If so, the UK could compete with Japan. BT has plans to use radio telephones in rural areas where cabling would not be economic.

Regional development

Swindon
Thorn EMI owns the cable in Swindon and has been given a licence to extend this for multichannel TV use. At the same time, the pilot plans for Cabletime/Water Research Council using fibre optics in the

sewers are taking place in this area. It will be interesting to see if there is any interaction and exchange of information between these two projects.

Westminster

Of all the areas given a licence, the interplay between leisure and the business use of cable will be more intense in this area. Therefore, careful monitoring of development in Westminster will be useful for future planning of communication systems.

Milton Keynes

Since its inception, Milton Keynes has been cabled by BT for domestic use. Non-interactive coaxial cable is used. This is placed in trunking, together with the telephone lines. Therefore, there are no aerials or telegraph poles in the new areas of Milton Keynes.

BT has wired up 18 households at Oldbrook, a grid square in Milton Keynes, with fibre-optic cable in switched star configuration to test the new cable topology which is now being employed in the latest UK cable systems. As there are access points to the trunking, it will be possible to wire up most of Milton Keynes with fibre optics as it becomes economically viable. BT has stated publicly that it regards Milton Keynes as a flagship for new technology and also has a commitment to this area because it plans to move its national network's headquarters there.

Acknowledgements

Much of the material in this paper was based on a regular newsletter for Aslib Information.

3: An information communication strategy — a case study

I G Dewis

BP International Ltd
London
UK

A strategy can never be a statement of fact – it is a signpost pointing in the direction of a recommended destination which may be many years away. It is important that each company should develop an information technology strategy to address its business needs and thus be in a position to make maximum use of the technical developments as they become appropriate. This case study details the strategy developed by a major British car manufacturer.

I G Dewis

Dr Dewis is Manager, Telecommunications Division, BP. Since graduating in Electrical Engineering in 1967, he has worked as Senior Scientific Officer at the Division of Computer Science, UK National Physical Laboratory; as Chief Technical Officer at British Steel Corporation; as Manager, Advanced Communications Systems, Development, and then Technical Strategy and Research at BL Systems (now Istel). Dr Dewis is a Chartered Engineer and a Fellow of the Institution of Electrical Engineers and of the British Computer Society. He is the author of many published papers on computer networks, packet-switching software, information communication and on standards in these areas, as well as on related management techniques.

An information communication strategy: a case study

Introduction

Mechanical switching systems and analogue transmission techniques are being replaced by digital switching and transmission. These will be achieved by British Telecom (BT), for example, through its System X family of advanced software-controlled systems and its X-stream services. It is estimated that optical fibre digital systems will account for half the BT transmission network trunk capacity by 1990.

Equipment manufacturers in the Information Technology (IT) industry are also taking great strides forward, but not always in the same direction. There has been very little natural compatibility between manufacturers except where one or two standards exist or where some manufacturers have decided to exploit a large market created by another.

Technological strategies

It is important that every company, whatever its size, should develop IT strategies to address its business needs and thus be in a position to make maximum use of technological developments as they become appropriate. Different business needs require different strategies and therefore it should not be assumed that the strategy detailed here is directly relevant to any other company, but experience gained at BL Systems Ltd, renamed Istel Ltd in January 1984, shows that all companies should be thinking of developing their own equivalent strategies.

Companies should not be tempted to exploit technology for technology's sake, but should use it only where it can be seen to support their business needs. Examples of such temptations are all too clearly seen in the area of office systems, where many companies are currently installing individual modules which do not form a consistent whole.

Information communication

Istel's information communication strategy is one of its IT strategies, but what exactly is 'information communication'? 'Information' is a global phrase used to encompass a wide range of traffic types including data (structured format, for example as output on a line printer in an EDP environment), text (unstructured, including telex, teletex and communicating word processor traffic), image (eg public and private videotex service, facsimile, etc) and speech. 'Communication' is composed of three parts — transmission, switching and value-added functions. Transmission is the transport of information from the output of the source device to the input to the destination device; switching is the ability to connect one device to a range of others; and value-added function is the ability of the network to offer a number of services useful to the user, eg conversion of speed, character code, protocol, storage, integrity, privacy, security, gateways, etc. In this paper, the first of these three is referred to as 'interlinking' and the second two (switching and value-added function) are referred to as 'interworking'.

Standards

As has been stated earlier, the problem in the information systems world is that interlinking and interworking standards tend to be set by manufacturers to cover their own products. Some standards exist on interlinking but standards for interworking are virtually non-existent. This problem is being addressed internationally through various standards organisations and the Istel strategy follows the route being established by the International Standards Organisation (ISO) with its Open Systems Interconnection (OSI) recommendations.

Istel

Istel is a wholly-owned subsidiary of BL plc. It offers systems and communications services to the parent company which is its major customer, although it is expanding its external customer base. The company operates a large communications and data centre in Redditch, Worcestershire, comprising mainly IBM and IBM-compatible equipment. Computers are also widely used in the area of production and process control at plant level where equipment mainly from Digital Equipment Corporation (DEC) is used. Equipment compatible with the communications standards set by IBM and DEC represents about 90% of the installed base of information systems equipment within the company and provided that users follow the Istel procurement policy, this situation will continue.

Istel's information communication strategy was developed by its former Advanced Systems Department — a research and development function within the company with the brief to develop and maintain strategies in key areas, to evaluate new technologies and methods and back these with pilot projects where necessary. In asking the Advanced Systems Department to develop and implement such strategies, the company recognised the need and importance of these strategies and has already derived great benefits from their existence.

Pivotal strategies

The IT strategies of Istel focus on three key or pivotal strategies: data, processing and communication. Communication is sometimes believed to be the dependent variable but in reality it is much more than this — it is an opportunity. A company that gets its communication strategy right will find that it can do much more with data and processing. Communication is the 'glue' which holds systems together and it is therefore imperative that all aspects of communication are given the attention they merit.

These three pivotal strategies (Figure 1) in turn lead to other strategies. Technical substrategies are the more detailed statements of the relevant key strategies. Supportive strategies enable the company to enact the key strategies and these include procurement, productivity and training. As far as the user is concerned, however, these strategies combine to produce the most important set of strategies — application strategies — which in Istel's case include factory, office, engineering and administration strategies.

Information communication strategy

Istel's information communication strategy was developed during 1981 and approved at the end of that year, after which its implementation commenced. Figure 2 illustrates the developments within this strategy. They are divided into two phases: phase one, where analogue transmission will continue to be used until some time in the 1990s, and phase two, where digital techniques will be exploited. Phase one is still active and the second phase has now commenced with digital transmission being increasingly adopted as its use becomes more cost effective. The two phases thus overlap for several years, ensuring a smooth conversion from analogue to digital working.

Istel's key activities for development within the two elements of interlinking and interworking are given below. As a guide, interlinking equates generally to levels 1, 2 and 3 of the ISO OSI model and interworking equates to levels 4-7.

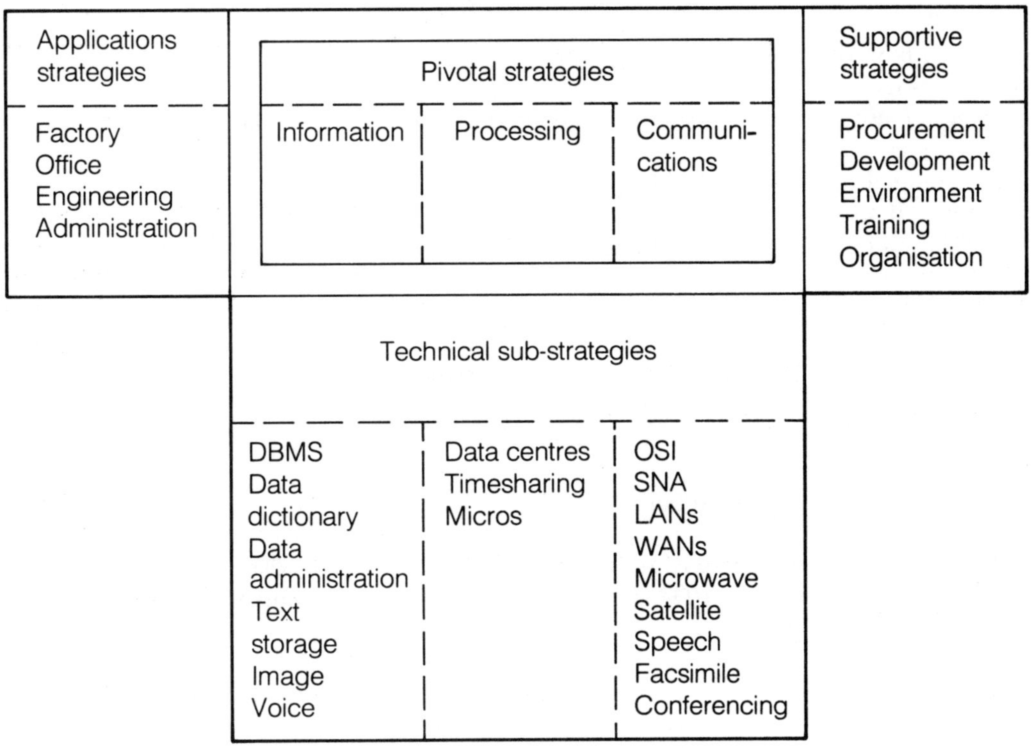

Figure 1: Strategies for information technology

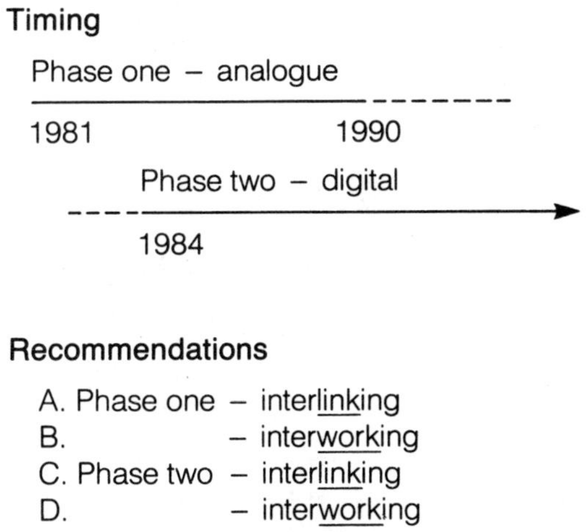

Figure 2: Strategy – key points

Phase one — interlinking

The aims of the phase one interlinking programme are:
- To remain analogue
- To make better use of existing facilities, having a period of consolidation to maximise the use of existing resources
- To force a minimal integration of non-speech traffic (ie data, text, image)
- No integration of speech and non-speech traffic
- To adopt OSI interlinking protocols in place of proprietary ones.

Phase one — interworking

Likewise, the aims of the phase one interworking programme are:
- To adopt OSI interworking protocols in place of proprietary ones
- A three-stage approach to be adopted to achieve this (see Figure 3).

The three-stage approach is outlined below:

1 Stage 1, which lasted from late 1981 until the middle of 1983, used existing IBM protocols to link the three generic types of system — IBM, DEC and $\propto$, where $\propto$ represents the internally designed office systems architecture.

2 Stage 2, from about 1983 to 1986, still exploits IBM protocols from IBM to DEC and from IBM to $\propto$, but gradually introduces OSI protocols between DEC and $\propto$.

3 Stage 3, from about 1986 onwards, will gradually replace proprietary communication protocols with their OSI equivalents where these are seen to support the business needs of Istel and its customers.

Phase two — interlinking

The aims of the phase two interlinking programme (see Figure 4) are as follows:
- To migrate gradually from analogue to digital transmission
- To develop a wide area digital network to offer a cost-effective inter-site facility
- To develop LANs at selected sites to provide cost-effective intra-site facilities
- To provide service networks on top of the digital network which each user or application will see, such that the basic digital infrastructure is hidden.

Phase two — interworking

Finally, the aims of the phase two interworking programme (see Figure 5) are as follows:
- To develop applications to use the most appropriate service network
- To continue implementation of OSI interworking protocols
- To place emphasis on supporting internal users when accessing external services and on external users accessing Istel's services via gateway devices.

Summary

Phase one concentrates on efficient interworking utilising existing communication facilities, whereas phase two is a gradual move towards digital transmission and the adoption of OSI protocols, providing a smooth introduction of new communication techniques.

Additional technical substrategies need to be developed in order to enhance and enact the information communication strategy. The areas included at Istel are IBM's System Network Architecture (SNA), DEC's DNA Local Area Networking (LAN), Wide Area Networking (WAN), exploitation of Istel's microwave network, as well as speech, facsimile, teleconferencing and the potential use of satellite capacity. These are further listed in Figure 1.

OSI

In this strategy, strong emphasis has been placed on OSI protocols. Istel's strategic direction is that all

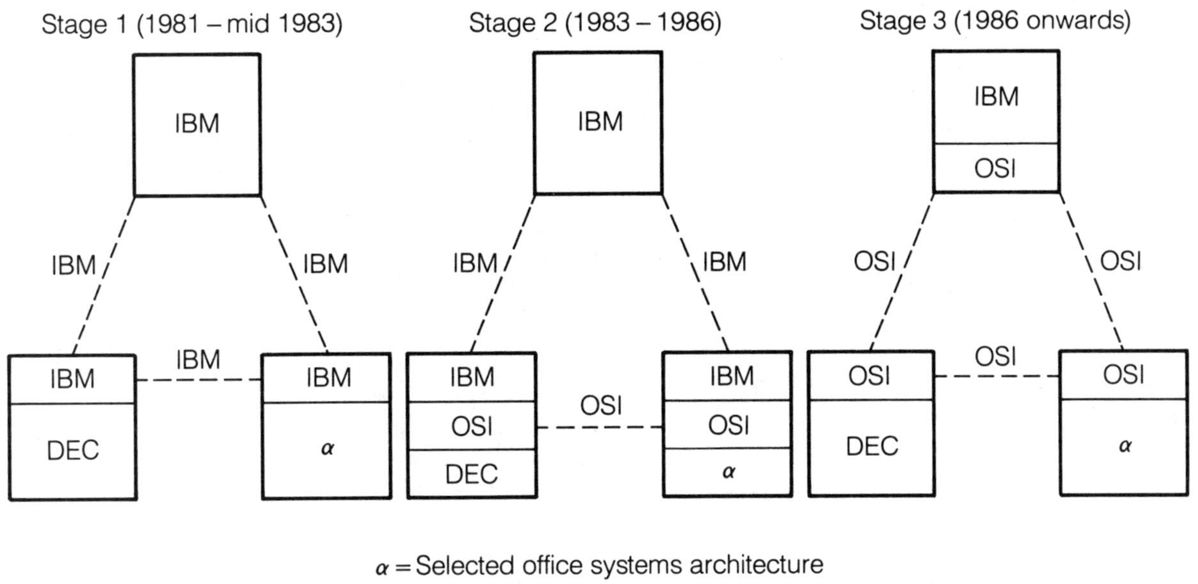

Figure 3: Three-stage approach to adoption of OSI protocols

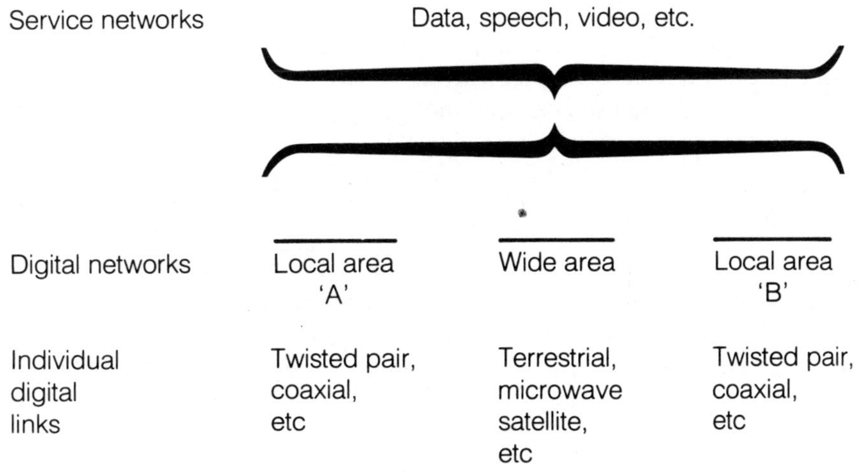

Figure 4: Hierarchy of digital network layers and connected services

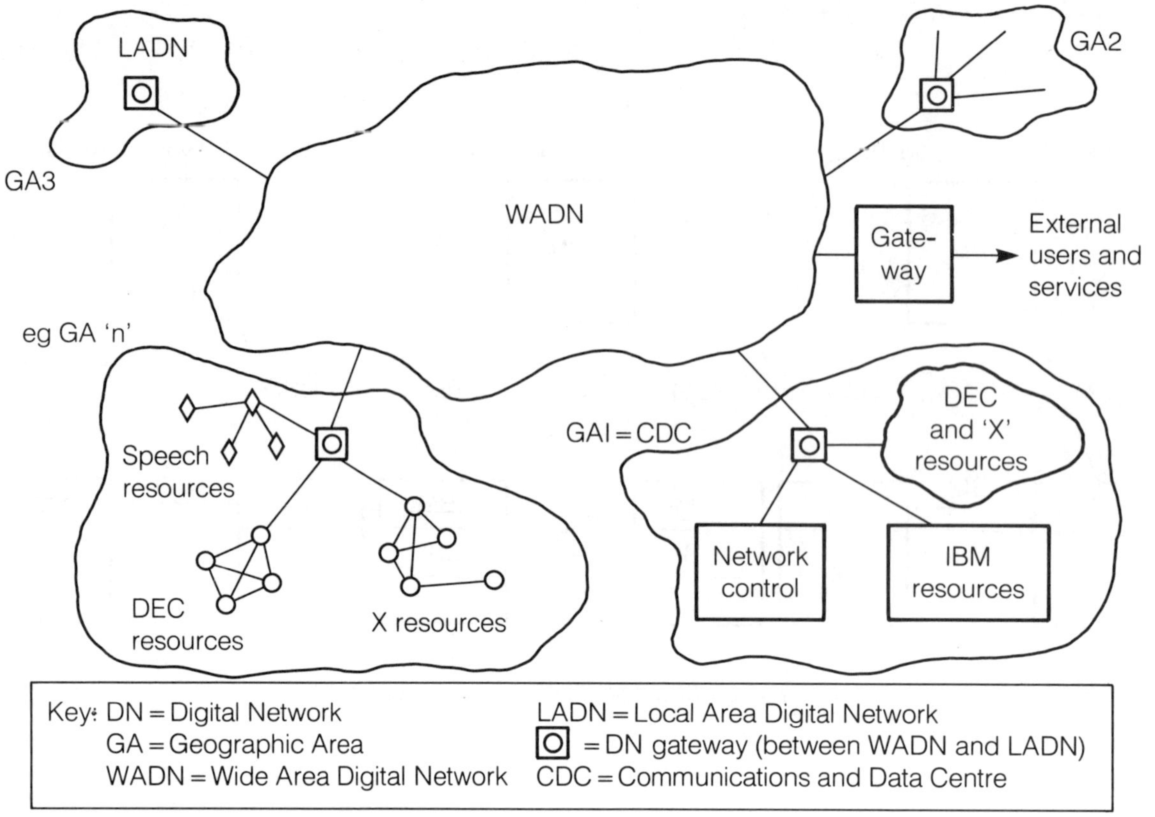

Figure 5: Example of integration of information communication exploiting digital infrastructure

applicable OSI protocols will be adopted, both for interlinking and for interworking. As a deliberate policy, however, Istel will not develop its own OSI protocol software package — it will only take them from manufacturers as they are offered and are fully supported.

In reality, the information communication strategy may never be fully achieved if full OSI protocol software is not made available. Nevertheless, any inroads made in this area will be tremendously beneficial. Whereas IBM and DEC equipment currently forms the mainstay of the procurement strategy, once OSI standards are achieved it will open up the way to bring in equipment from other manufacturers much more easily. In those areas where OSI interfacing software does not become available, it is likely that low-cost microprocessors will be used as emulation and translation devices.

Conclusions

A strategy can never be a statement of fact — it is a signpost pointing the direction towards a recommended destination which may be many years away. The installation of systems which form part of a strategy and which will only 'come together' in the future can cause problems for systems implementors who want to use today's proven technology and install something that they know works now. It is obvious that a strategy cannot dictate the precise timescales and detailed tactical steps required to achieve the end point — these must be determined on an 'as and when' basis.

A strategy may never be accomplished totally. Views may be modified based on two factors — changing technology resulting in new or more cost-effective facilities and changing end-user requirements which can be affected by user awareness of systems capabilities. Either of these could render parts of the strategy inapplicable. Consequently, Istel believes that its key strategies will need to be reviewed and perhaps refined every two or three years.

4: Transmission media of the future

J E Ettinger

Standard Telecommunications
Laboratories Ltd
Harlow
Essex
UK

When studying the possible transmission media of the future, it is essential to consider not only the introduction of new media, but also the ways in which existing media are likely to be utilised in the future. The next 20 years pose a formidable challenge to the transmission technologists. The evolution of new services is discussed in this paper and their advantages and disadvantages are highlighted.

J E Ettinger

After gaining his first degree in Electrical Engineering at the University of London, Mr Ettinger took a graduate apprenticeship and worked as a systems planning engineer in the communications industry specialising in non-voice communications. As a Senior Principal Engineer with Marconi Space and Defence Systems Ltd he was concerned with design of total communications systems for Command Control Communications. At present he is a Senior Principal Research Engineer in the Systems Planning Division of Standard Telecommunications Laboratories Ltd (STL) developing integrated services, digital networks and local area networks. He has participated in the work of the CCITT and other similar bodies on data communications and teleprocessing.

Transmission media of the future

General overview

The modern telecommunication network allows its users to connect some 500 million users to each other: the parties may be next door to one another or many thousands of miles apart. This remarkable world-wide system is the end-product of over 1000 billion dollars (at 1983 prices) of accumulated investment in the last 20 years by the telecommunications carriers all over the world and the willingness of these carriers to cooperate with each other in matters of tariffs and standardisation.

The telecommunication network may be visualised as a hierarchy of transmission links interconnected by switching centres. Subscribers gain access to their local exchange by means of the local distribution system (local loop or local network), using the signalling arrangements that control the operation of the network. The network hierarchy has four main parts:

- The local distribution part of the network, connecting the individual subscriber's terminal equipment to the local exchange
- The links that connect nearby local exchanges to one another and local exchanges to trunk access exchanges (junction network)
- The long-distance network that provides all other interexchange links within the country (trunk network)
- The international network.

A brief examination of the capital expenditures shows that about half of all investment by the telecommunications carriers is concentrated on the transmission components of the system. There are, however, some wide disparities in the cost mix between the local and the long-distance parts of the network. In the long-distance network, transmission costs dominate the cost equation, whereas for the local networks switching and signalling costs are dominant. The introduction of optical fibre technology into the local distribution may well change this mix of costs, although only the very earliest stages of this process can yet be observed in practice.

Since the costs for the shorter-distance traffic system components other than transmission (local distribution, switching and multiplexing equipment) are not falling rapidly, while the long-distance transmission costs are, the end result has been substantial relative reductions in the price of long-distance and international calls. These gains reflect three main sources of cost reduction in transmission: the introduction of new technologies (such as satellites and long-haul optical fibre systems), the continuing rapid improvement of the 'mature' technologies (such as microwave radio and coaxial cable) and the improved utilisation of the existing plant. For example, a single pair of wires in a multipair junction cable linking two local exchanges and initially carrying a single analogue signal can with digital transmission techniques — specifically Pulse Code Modulation (PCM) — carry numerous one-way voice or data channels. When considering the transmission media of the future, it is essential to consider not only the introduction of the new media, but also the way that existing media are likely to be utilised in the future.

The local network and the twisted pairs

It must be recognised that the local distribution network and the junction network account for substantial proportions of the total capital expenditure by the telecommunications carriers. It is, therefore, highly unlikely that this investment will be abandoned, although it may be supplemented by the techniques emerging from the new technologies, primarily coaxial cable and optical fibre systems.

The local and the junction networks generally connect the subscriber terminal equipment to the local exchange and link the local exchanges with each other, via circuits comprising two wires twisted together — the so-called twisted pair. In the early days, voice transmission followed the pattern established for telegraph lines, using open wire lines supported by poles. However, as the frequencies used in telephony are significantly higher than in telegraphy, the attenuation of the telephone signal was greater than for the telegraphy service. The technological effort was thus concentrated on extending the distances over which the signal could be transmitted. This resulted in the use of copper instead of steel and, due to the developments in the handset technology, reduction in the actual amount of copper used in the local network.

The current trend is the gradual introduction of the multiplexing techniques at low levels of the hierarchy in the junction network. This trend began approximately 40 years ago with the introduction of the Frequency Division Multiplexing (FDM) techniques into the local networks capable of carrying the equivalent of 12 to 24 full-duplex voice channels over two twisted pairs of wires. The trend is continuing, using the digital transmission techniques at 2 Mbit/sec in the junction networks using cable systems.

The introduction of the digital, Time Division Multiplex (TDM) PCM techniques in subscriber networks will not only lead to the lowering of costs but may also open up many new service options. The twisted pair can, in principle, carry data at rates up to 144 Kbit/sec. This allows for the introduction of digital services such as teletex, videotex as well as voice, right down to the subscriber terminal equipment. These services may also include data transmission, electronic mail, teleconferencing and electronic fund transfer, which will be carried in accordance with a verified set of digital standards, within an Integrated Services Digital Network (ISDN).

Coaxial cable

One major focus of research in the transmission technology has been the search for new media capable of carrying ever larger numbers of signals. The coaxial cable, with its self-shielding properties was the first transmission medium where this search yielded major gains in transmission capacity. A coaxial cable is a transmission line made up of an inner conductor surrounded by a concentric earthed outer conductor, separated by an insulator called the dielectric, consisting of different kinds of materials, eg polyethelene, gas, air.

Developments in coaxial cable technology have concentrated on increasing the operating bandwidth of the cable. Currently such cables can carry up to 108 000 channels. More recent research has concentrated on the application of the coaxial cable to the so-called cable TV systems. In this application, in Britain, they can carry up to 30 TV channels.

The chief disadvantage of coaxial cable systems is that they require frequent use of signal repeaters, the spacing of which is dependent upon the carrier frequency and the transmitted bandwidth. For example, to carry four TV channels on UHF, the required repeater spacing would be approximately 300 metres.

Optical fibre systems

The use of light for the transmission of information is far from being a new idea. It was Alexander Graham Bell's experiments more than a century ago that led to his development of the photophone, a device that carried speech from one point to another by means of vibrating mirrors and a beam of sunlight. Although not a commercial success it nevertheless demonstrated the feasibility of the light wave communications. However, it would very likely have remained in limbo had it not been for the appearance of a new device called the laser, which is capable of producing the so-called coherent light, which has the properties of emitting light waves that are of the same wavelength, are in phase and hence can be sharply focused to travel in the same direction over long distances with very little loss of power, although the laser used in the fibre system differs from those employed for communication in 'free' spaces.

The concept behind fibre optics is simple. In the same way that a twisted pair can carry an electrical signal, the optical fibre carries a light signal. However, fibre optics offers some major advantages over all other transmission media:

1 The basic raw material is sand, the most common material in the world.

2 Its physical size (approximately one-tenth of a millimetre in diameter) is much smaller than that of copper wire; hence, greater numbers of fibres can be packed in a small volume with subsequent savings in capital and installation costs.

3 Attenuation is low. The current state of technology permits the repeater spacing at approximately 30 km apart. As the technology matures, this distance is likely to increase. This implies savings in remote power feed systems, civil works, etc.

4 Since it operates using light signals, the system neither generates electrical interference, nor is adversely affected by it.

5 The medium has substantial transmission capacity, since the frequencies of the order of 10^{15}, some 10 000 times those used for microwave, are used for transmission

Optical fibre systems consist of a light source, the optical fibre, the cable containing the fibre, a light detector and the connectors and couplers needed to assemble the above components. The light signal is modulated with information derived from the established electronic equipment to drive relevant electro-optical equipment.

Fibres

In order to understand fibre optics more fully, it is well to know how light behaves when it travels through matter. All materials through which light can travel are classified by their index of refraction, the ratio of speed of light in vacuum to its speed in that particular material. Light waves of the same frequency travel through different materials at different speeds. A light ray bends as it passes through one transparent medium to another. This phenomenon is known as refraction. When light passes from a medium with a higher refractive index to one with a lower one, the ray is bent. At a critical angle the light will not pass from the medium with a high index to one with a lower index, but it will be totally reflected into the original medium.

In optical fibre systems, light is directed at a low angle into a fibre consisting of one transparent material — the core — which is surrounded by another material having a lower refractive index — the cladding. The light then remains within the fibre and is reflected back and forth along the interface of the two materials.

The early optical fibres had attenuation, or losses, of approximately 20 dB/km. The modern fibres, however, are now available with losses of approximately 0.5 dB/km.

A second technological obstacle that had to be overcome in developing practical fibre systems was the 'dispersion' of light as it travelled through the fibre. This is caused by the fact that light consists of a range of wavelengths which travel at different speeds. This phenomenon limits the bandwidth capacity of optical fibre systems. Also, the light waves tend to spread as they travel, a fact that is exaggerated by the additional dispersion occurring within the fibre because the beam from the light source consists of rays which are not absolutely parallel to the central axis of the fibre.

This form of dispersion, called modal dispersion, severely limited the information capacity of the early fibres, the so-called step-index fibres, the core of which had one refractive index and the cladding another. Modal dispersion in these fibres led to development of two other types — graded-index and single-mode (monomode) fibres.

In graded-index fibres, the refractive index varies across the cross-section of the glass. It is higher at the centre than it is at the outer edge. This has the effect that the light beam is continuously refocused as it travels down the fibre, thus keeping the dispersion low.

In single-mode fibres the central core is made so fine that only a very narrow beam from a single-mode laser can pass through it with minimal dispersion. Single-mode fibres are likely to be the best choice for long-distance transmission in the future.

Light sources

Only two kinds of light source, both based on semiconductor technology, are appropriate for fibre optics: the Light Emitting Diode (LED) and the semiconductor laser. LEDs are less sensitive to power fluctuations and to changes in temperature and humidity. On the other hand, lasers emit a light beam of high intensity of narrower spectral width (20 Angstroms for lasers; 350 Angstroms for LEDs) thus giving less material dispersion.

Detectors

Two types of semiconductor detector can be used in fibre-optic systems: the p-i-n photodiode and the avalanche photodiode. The former is simpler and it can be used with LED sources and the medium-to-high loss fibres. The avalanche photodiode is more complex, is more efficient across a wider spectrum, suffers less from interference and has a faster response time to signals than the p-i-n photodiode.

Connectors and couplers

Due to the precision with which the two adjoining fibres must be joined, special joining and coupling techniques have been developed. These involve the use of a short-pulse plasma arc to join the two ends of the fibre together which have been very accurately positioned beforehand by the use of special equipment.

Integration

Work is now progressing on the development of Integrated Optical Circuits (IOC) consisting of a source, fibre and detector, integrally mounted to create a very small unit able to perform all the functions of a conventional integrated circuit but at faster speeds and eliminating the need for some of the optical-to-electronic signal conversion.

Applications

Applications of fibre-optic systems are likely to be found in intercontinental submarine systems, long-haul land transmission, cable TV systems, LANs and short-haul transmission particularly under high interference conditions, eg telemetry inside power stations, coal mines, oil fields.

The major challenge currently facing optical fibre systems technology is the fact that the present systems are power operated. There is no equivalent of a high-impedance tap. Other challenges lie in the cost reduction of optoelectronics and the development of wavelength division multiplexing to derive multiple, unidirectional transmission channels on a single fibre.

The other major challenge is presented by exploitation of coherent techniques to increase receiver sensitivity and thus maximise repeater spacing.

Radio systems

Line-of-sight microwave systems

Above about 1000 MHz, radio waves travel predominantly in straight lines and are attenuated due to the 'free-space' losses. Signals must, therefore, be transmitted from one receiving tower to another along the line-of-sight. These attenuation and directional properties have some advantages because, unlike radio systems operating at low frequencies, these microwave frequencies can be used in relatively small areas.

Technological improvements in line-of-sight microwave transmission have been concentrated on increasing the channel capacity and directional sensitivity of the antenna and, more recently, improving the design of digital radio transmission systems. Line-of-sight links are usually found on high-capacity medium-haul routes, being highly attractive where terrain would present major problems for cable transmission systems, and for any temporary high capacity links. Their capacities compare with those encountered in cable transmission systems and their application has been found appropriate not only to the transmission carriers for long-haul applications, but also by broadcasting authorities for short-haul transmissions and for links to the local nodes and others such as the proposed Mercury carrier.

Tropospheric scatter systems

Microwave frequencies used for line-of-sight communications have fixed capital associated with providing repeater stations at approximately 50 km intervals. They are, therefore, unsuited where traffic capacities are low or where such repeater sites cannot be found. An alternate technique, known as tropospheric scattering is used in such cases. This technique allows for over-the-horizon transmission using the scatter properties of the troposphere. Distances of up to 450 miles can be traversed with one jump using these techniques. Such systems do tend to have high power requirements and need large, complex antenna systems. Normally, this type of system is viable only for links with relatively small traffic volumes, over difficult terrain, over water (eg offshore oil rigs) or in underdeveloped areas. They are rapidly being made obsolete by the decreasing cost of competing satellite technology.

Cellular radio systems

Another major innovation that is well advanced in development is that of land mobile services. Mobile radio telephone services are not a new concept, being currently available on a commercial basis in most industrialised countries. Demand for such mobile services has, however, always outstripped capacity. Present-day mobile telephone services use a high-powered radio transmitter with a 2500 square mile coverage. Each transmitter normally has about 12 channels, resulting in severe congestion problems despite limitations on the number of mobile telephones.

The concept of a cellular mobile radio system overcomes this capacity problem. Instead of one high-powered transmitter, the region is broken up into smaller areas — called cells — of one to eight square miles, each containing a low-power transmitter. The low-power transmitters can communicate with the mobile telephones via radio and are further connected to network switching or control centres through the telephone wires. Adjacent cells use different frequencies to ensure that there is no interference between cells. Each area is assigned about 300 voice channels in the UHF frequency range.

The idea behind this cellular band mobile telephone system is simple. As the mobile telephone travels through a cell, it broadcasts its presence to transmitters capable of receiving the signal. The strength of the signal is used by the control centre to determine the transmitter closest to the mobile telephone. As the mobile unit moves out of an area, the control centre dynamically reassigns the transmitter and informs the mobile telephone equipment that another carrier frequency should now be used to continue the conversation. The mobile telephone equipment switches over to this new frequency and is thus in contact with a closer transmitter. The process of such a handover between two transmitters is achieved automatically and very rapidly so that the user does not notice the change. Driving out of the control centre's area is also no problem, since when this happens the adjacent control centre informs the previous one, and all calls are routed to this new area.

This cellular radio telephone network may in the future be used not only for communication with the mobile subscribers, but also telephone administrations could use this technique as a quick and easy method to provide temporary installations that are often required when major events are held in a particular geographic area.

Satellite systems

The launch of the Early Bird satellite in 1965 heralded the beginning of the space age in respect of communication satellites, revolutionising the means of transmission of voice TV and data between the nations. This has now been extended to include mobile applications both in the military and commercial fields.

Satellites, which in many respects are really radio relays in space, enable point-to-point and point-to-multipoint connections to be made under circumstances where terrestrial means may be impracticable, eg in vast expanses of territory where population centres are well scattered or the country is generally sparsely populated as in countries like Canada and the Soviet Union. The most spectacular use of satellites is in the way they enable live news coverage of events in remote parts of the world to be achieved or make available news pictures of an event the day it has taken place. A further feature is the use of satellites for the relay of TV programmes via satellite from a central unit to numerous well-scattered ground transmitters which finally broadcast the picture to the viewer. Now direct broadcasting of TV to the home is being planned in a number of countries, starting in 1985.

Intercontinental systems

The first application of satellites was on an intercontinental basis with the launch of the Intelsat range of spacecraft under the auspices of the International Telecommunication Satellite Consortium created in 1964. It has been responsible for international cooperation in the design and operation of a range of spacecraft for the transmission of wideband signals. The current spacecraft is Intelsat V; Intelsat VI is under manufacture. Three spacecraft in a geostationary orbit give global cover with the exception of the high latitudes near the poles. If polar coverage is required a different orbit must be used, eg the Soviet Union uses a highly elliptical polar orbit to enable its northernmost areas to be covered.

Intelsat has generally used analogue transmission using Frequency Division Multiplexing (FDM) in its satellite systems but the trend is now towards digital transmission in the form of Time Division Multiple Access (TDMA) and this will be the dominant feature of Intelsat VI. The European Communication Satellite (ECS), which was launched in 1983, is fully digital and can handle digit rates up to 120 Mbyte/sec.

Most spacecraft carry a number of transponders, each of which is capable of carrying the equivalent of about 960 telephone channels or one/two TV channels. The total equivalent capacities of some spacecraft are given below:

- Intelsat IV — equivalent number of telephone channels (half circuits): 10 000
- Intelsat V — equivalent number of telephone channels (half circuits): 25 000
- Intelsat VI — equivalent number of telephone channels (half circuits): 30 000.

Most communication satellites to date have operated in the 6 and 4 GHz radio frequency bands for their up and down transmissions respectively. Current Intelsat satellites and ECS use the 14 and 12 GHz bands and the effect of rain attenuation has to be taken into account in the design of the system. Earthstation standards regarding receiving sensitivities, the so-called figure of merit, have been established by Intelsat for its system and all potential users must comply with these standards before they can be integrated into the system. The larger Intelsat earthstation has an antenna of around 30 metres diameter while the smaller ones use a 10-metre diameter antenna. The charge made by Intelsat for the use of its space segment is inversely proportional to the receiving sensitivity of the earthstation concerned. Some users prefer to save on the capital cost of establishing an earthstation and pay increased rentals to Intelsat instead.

The majority of spacecraft are located on the geostationary orbit and some parts of this orbit are more likely to become congested than others, eg the part of the orbit which is of interest to users from North and South America. The question of orbit allocation and the spectrum usage in the geostationary orbit is under active study by the International Telecommunications Union.

Domestic satellite systems

Satellites have been used for domestic systems for some years in Canada, the US and the Soviet Union. The European system should also be included in this category since it is not an intercontinental system. As already mentioned, some of these systems are used for TV distribution but they are also used for business purposes, particularly in the US. In Europe the current ECS and the planned L-Sat (launch date 1986) are the main systems but there is also Telecom 1, a French satellite system, which will be used for the transmission of digital traffic.

Mobile systems

In parallel with Intelsat, a similar international organisation is concerned with the use of satellites for communication with ships. This is the International Maritime Satellite Organisation (Inmarsat) which includes 26 signatories. It controls the operation of the system in the three principal oceanic areas of the world, ie Atlantic, Pacific and Indian Oceans. In mid-1983 there were 1850 users of this system, which includes ships, offshore drilling rigs and static oil platforms.

The Inmarsat system is an excellent example of international cooperation in that it consists of three Marisat (US) spacecraft, one Marecs A (European) and one maritime package on an Intelsat V satellite. These spacecraft use either the 4/6 GHz or the 12/14 GHz bands for the path from the shore station to the spacecraft, but all use the same 1.6 GHz band for the path between the spacecraft and the ship. The maximum number of telephone channels, in the form of Single Channel Per Carrier (SCPC) per spacecraft is now in excess of 50.

Aerosat systems have been under discussion, off and on, for a number of years but the airline industry is not convinced that the improved reliability of transmissions made available by the use of satellites justifies the substantial increase in cost in providing these communication facilities.

Direct broadcasting systems

Direct broadcasting from satellites has been available for some time for the user having a sufficiently large receiving antenna. An example of this is the ECS satellite for the transmission of TV in Europe during 1984. True direct broadcasting to the home, for which antennas of around 1-metre diameter will be required, will begin in Europe in 1985 with the Franco-German TDF1 satellite, followed in 1986 by both UNISAT (the UK satellite) and L-Sat (the ESA European satellite). Some eight organisations in the US are planning to launch direct broadcasting satellites in the 1980s. Other countries, eg Japan, India and others in South America, are planning to launch direct broadcasting satellites.

Currently planned broadcasting satellites will use the 12 GHz band for downpath transmission, with the carrier being frequency modulated. The up-path will be in the 14 GHz band.

Conclusions

The past 70 years have seen remarkable developments in transmission technology that have resulted in a manifold reduction in costs per voice channel for long distance calls. The primary factor behind this success in the lowering of costs has been the development of systems of higher and higher capacity using terrestrial cable — coaxial or optical fibre, radio and satellite systems. In the terrestrial systems, the operating characteristics are dominated by costs of land, buildings and installation. Subsequent developments that better utilise the given medium thus require only moderate increases in the cost of electronics, but enable the heavy fixed expenses to be divided among a greater number of voice channel circuits.

A large proportion of traffic will continue to be carried by the terrestrial transmission media as these economies of scale continue. Satellite communication technology will play a growing role, especially on the medium- to long-haul routes, as it gradually becomes more competitive over progressively shorter distances and wherever traffic densities are not high enough to exploit the economies of scale in the terrestrial transmission technologies. Thus, the terrestrial and the satellite technologies can be expected to develop complementary values in which the mix of transmission media will depend upon the physical, economic and traffic characteristics of particular routes. Together, these two streams of technological development will open up extensive opportunities for the economic and efficient provision of a vastly increased range of telecommunications services.

Furthermore, digital operation will enhance the quality and features of the services offered. For example, low-cost data communications — and hence a variety of applications, ranging from electronic mail to home banking — will, in due course, be provided between subscribers at speeds of 64 Kbyte/sec or more where it is economically justified. Expanded use of video services will also gradually become more economic, as will the provision of a wide range of services to mobile terminals.

The next 20 years will pose a formidable challenge to the transmission technologists. The evolution of new services such as electronic mail and large-scale mobile radio telephone services, plus the move towards all-digital operation in the networks, point to an exciting future. The basic technologies for such services are already with us: as these mature, they will provide the key element of a telecommunications revolution that has the potential to transform the economic prospects and daily lives of us all over coming decades.

5: Integrated Services Digital Network (ISDN)

J E Ettinger

Standard Telecommunications
Laboratories Ltd
Harlow
Essex
UK

The present convergence of computing and telecommunications is creating a market for telecommunications services in general and for integrated services in particular. Although the user will no doubt be influenced by the tariff structure, he must have the final choice of which services he might wish to use. This paper reviews: the status of the various services that are likely to be provided by ISDN; the interworking with other networks; and future ISDN developments.

Integrated Services Digital Network (ISDN)

Introduction

The present telecommunications network has been shaped over the last 100 years by inventions in telegraphy, telephony and facsimile. With the exception of telegraphy, man's means of communication have, until recently, been predominantly analogue because they correspond with his primary communication senses ie his eyes and ears.

The conventional telephone network in most countries is therefore based upon analogue transmission and switching with a restricted signalling capability. By following this network within a given set of national boundaries, a full-duplex path of about 3 kHz takes from five to 25 seconds to set up. Furthermore, the characteristics of the communication channel thus established (primarily those of attenuation and group delay) are rather poorly defined because the human ear, in combination with the human brain, is not particularly sensitive to the quality of the communication channel. The human voice is still the dominant mode of communication. Consequently the analogue signal, which mirrors the human voice by modifying (or modulating) an electric current along a transmission path, is still the predominant transmission method in the world's telephone networks.

However, the demand for expanded communications services is reflected in the growing importance of information-related activities to the creation of new jobs and products. This in turn creates a demand for a much wider range of integrated telecommunications services. While analogue and digital signals are based on different principles, information can be transmitted in either form.

Analogue and digital technology

Through the technique called Pulse Code Modulation (PCM) analogue signals can be converted to digital form and transmitted as a bit stream on a conventional telephone line. As a digital format, PCM can be handled by digitally-controlled, digital switching equipment which rapidly routes speech, data, telex, or visual information simultaneously through the network.

Digital transmission via the telephone network is critical in integrated telecommunications services, the improved quality alone being a major advantage. Digital pulses can be regenerated in their original form, whereas analogue signals must be amplified along the transmission path together with any noise or distortion that may be present. The digital pulses arriving at the final destination are the same as the original signals and they can be converted back into voice, written, or visual information.

An all-digital network addresses a variety of traffic characteristics eg for voice communication, the digital network accommodates real-time conversation more economically while at the same time, for non-telephone service transmission, the digital system can manage the traffic volume more efficiently through electronic storage or by changing the transmission speeds. Such efficiency is not possible in analogue networks where the ability to accommodate non-telephone services (such as telex) quite often involves the creation of separate networks, and requires different types of equipment.

Integrated Digital Network (IDN)

A fully digital network is referred to as an Integrated Digital Network (IDN). Its importance lies in the fact that once an analogue signal from a subscriber has been converted to a digital signal it remains digital for its entire passage through the network until it is finally converted back to analogue for application by the called party.

The introduction of Stored Program Control (SPC) and Common Channel Signalling (CCS) greatly enhanced the capability of a network to support advanced features, eg for new supplementary services (such as Closed User Group and Originating and/or Terminating line identity), operation and maintenance functions.

IDN is a particularly suitable vehicle for service integration since all types of information, including network control information, may be digitally encoded and then transferred as streams of bits which can be handled in a uniform way. Such an ISDN is expected to evolve from the telephony IDN by integration with other non-voice services. The general schematic of ISDN, as applicable in the UK, is shown in Figure 1.

In addition to the technical and economic arguments for ISDN, the present convergence of computing and telecommunications is creating a market for telecommunication services in general, and for integrated services in particular. The demand for integrated services arises from users' needs for telecommunication installation offering flexible access eg the Integrated Services Private Branch Exchange (ISPBX) and the Local Area Network (LAN).

A useful method of providing digital services, especially for business customers, is by means of Satellite Multiservice Systems (SMS). In the short term a satellite system with small earth stations close to users' premises could provide digital services until a terrestrial ISDN is available. In the long term such satellite systems, in parallel with a terrestrial ISDN, could provide services which benefit from special satellite system characteristics such as insensitivity of the cost to the distances to be covered, broadcast and point-to-multipoint services, and bandwidth.

Definition of ISDN

The current concept of an ISDN is based upon the I-Series Draft Recommendations of the CCITT Study Group XVIII; in particular, I-120 defines ISDN, as summarised below.

> *A network evolved from the telephony IDN that provides end-to-end digital connectivity to support a wide range of services, including voice and non-voice services, to which users have access by a limited set of standard multipurpose customer interfaces.*

This definition covers not only an 'ideal ISDN' where all integrated services share the use of all network resources, but also cases of partly-shared use of these resources such as, for example, access to the packet-switched network. It follows that the minimum requirements are:

- A common customer access
- Network resources, mainly switching and transmission, which are shared by voice and non-voice services.

Although the user will no doubt be influenced by the tariff structure he must have the final choice of which services he might wish to use.

The basic conceptual principles of ISDN as outlined in I-Series Recommendations are that:

- It should evolve from the telephony IDN based upon 64 Kbit/sec channels
- It should have both circuit- and packet-switching capabilities from a family of digital user/network interfaces
- Where applicable, it should possess hybrid access capabilities.

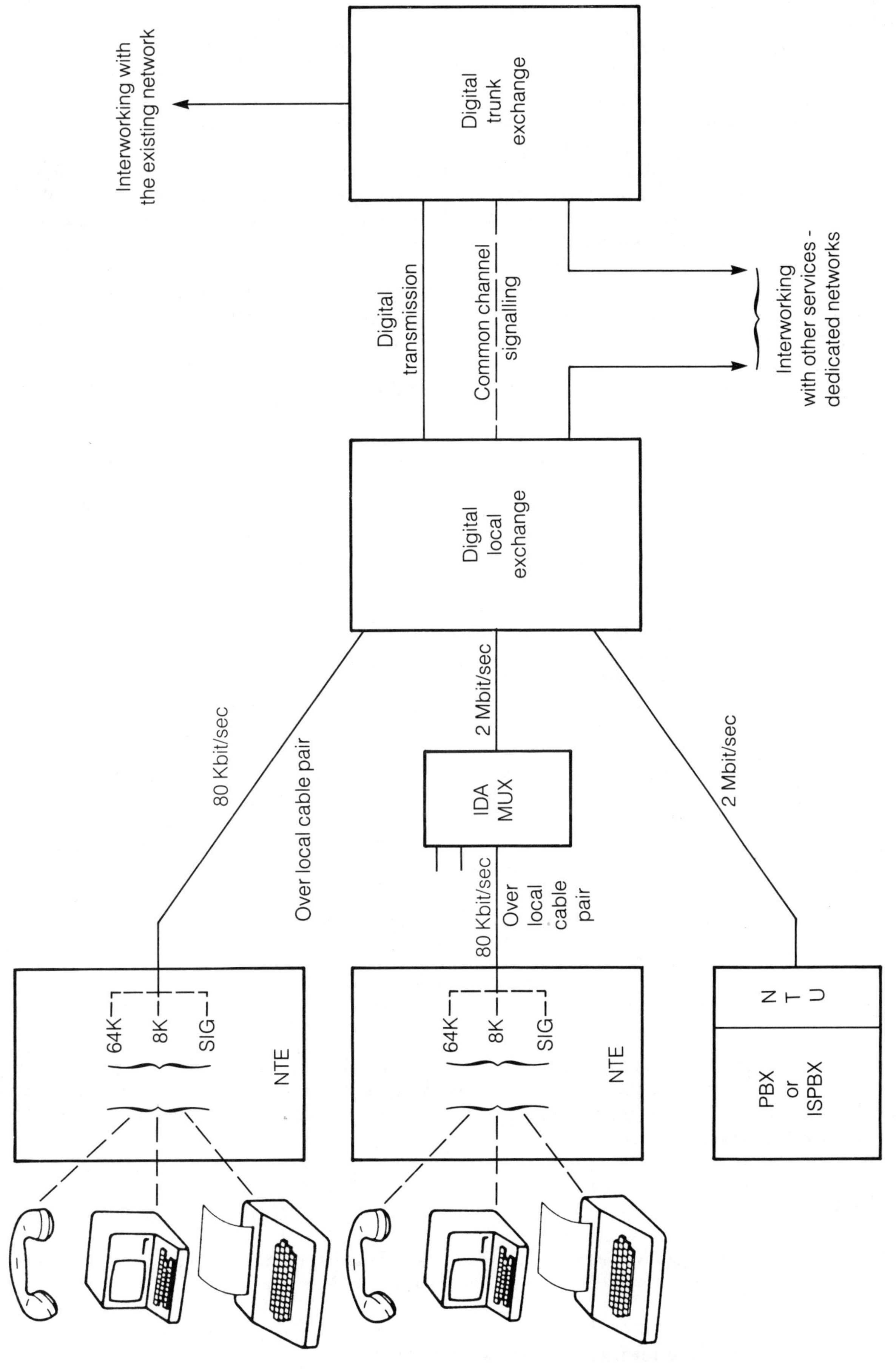

Figure 1: UK ISDN schematic

Basic structure

The general structure of an ISDN may be seen as consisting of a number of subsystems or component parts:

- Digital transmission and multiplex systems of various levels of the digital multiplex hierarchy
- SPC digital switching systems, primarily designed for switching of 64 Kbit/sec connections, but which may also include other features such as packet switching
- A common channel signalling network to interconnect all nodes of the communication network
- User access arrangements and interfaces to connect various user terminals, systems and networks to the ISDN. This access is, in principle, composed of one or more 64 Kbit/sec channels (or submultiples of that speed) and a message-oriented channel carrying information (including signalling information) associated with communication channels.
- Special network nodes dedicated to the provision of communication types different from the 64 Kbit/sec transparent connections such as packet switching, or the provision of centralised network capabilities for services with low penetration in the network
- Special nodes with gateway functions and interfaces for interworking with other dedicated networks
- Operation, maintenance, and operator assistance centres.

Access methods and signalling

Two types of access method will be open to the users (See Figure 2 for general arrangements in the UK.):

1 A single-line connection, known as the 'basic' access, which will provide for:
   - A connection of up to 64 Kbit/sec for either voice or non-voice use
   - Another connection for non-voice applications which may be less than 64 Kbit/sec (in different countries there are proposals for 64, 16 and 8 Kbit/sec)
   - A channel of 8 or 16 Kbit/sec which will carry signalling, protocol and rate adaptation information and possibly packet information
   - Access by means of Network Termination (NT) equipment located at the user's premises providing standard CCITT X or V interfaces.

2 For business users a multiline connection, known as the 'primary' access, will provide ISPBXs with up to 30 independent 64 Kbit/sec exchange connections carried on a 2 Mbit/sec digital path.

In both cases a common-channel, message-oriented signalling system will be used. In the UK this is based on the Digital Access Signalling System (DASS) which, at layer 2, is related to the ISO High-Level Data Link Control (HDLC) protocol. In other countries it is likely that a signalling system, known as LAP D, which is based upon the CCITT recommendations I441 and I451, may be extended to the user's ISPBX.

User services

It is planned that ISDN will support the telephony service, ISPBX and digital PBX access, private circuit data services, and circuit-switched services both synchronous (up to 64 Kbit/sec) and asynchronous (up to 9.6 Kbit/sec). It will also provide access to the Viewtex (or Videotex), eg British Telecom (BT) Prestel, and the packet-switched service.

ISDN in the UK

British Telecom has a huge programme underway to modernise its communications network. System X, which is central to this programme, also forms the heart of ISDN.

User/network interface and access arrangements

In order to provide a set of unified interfaces, British Telecom, the UK PTT, will supply two types of NTE at the user's premises (see Figure 3):

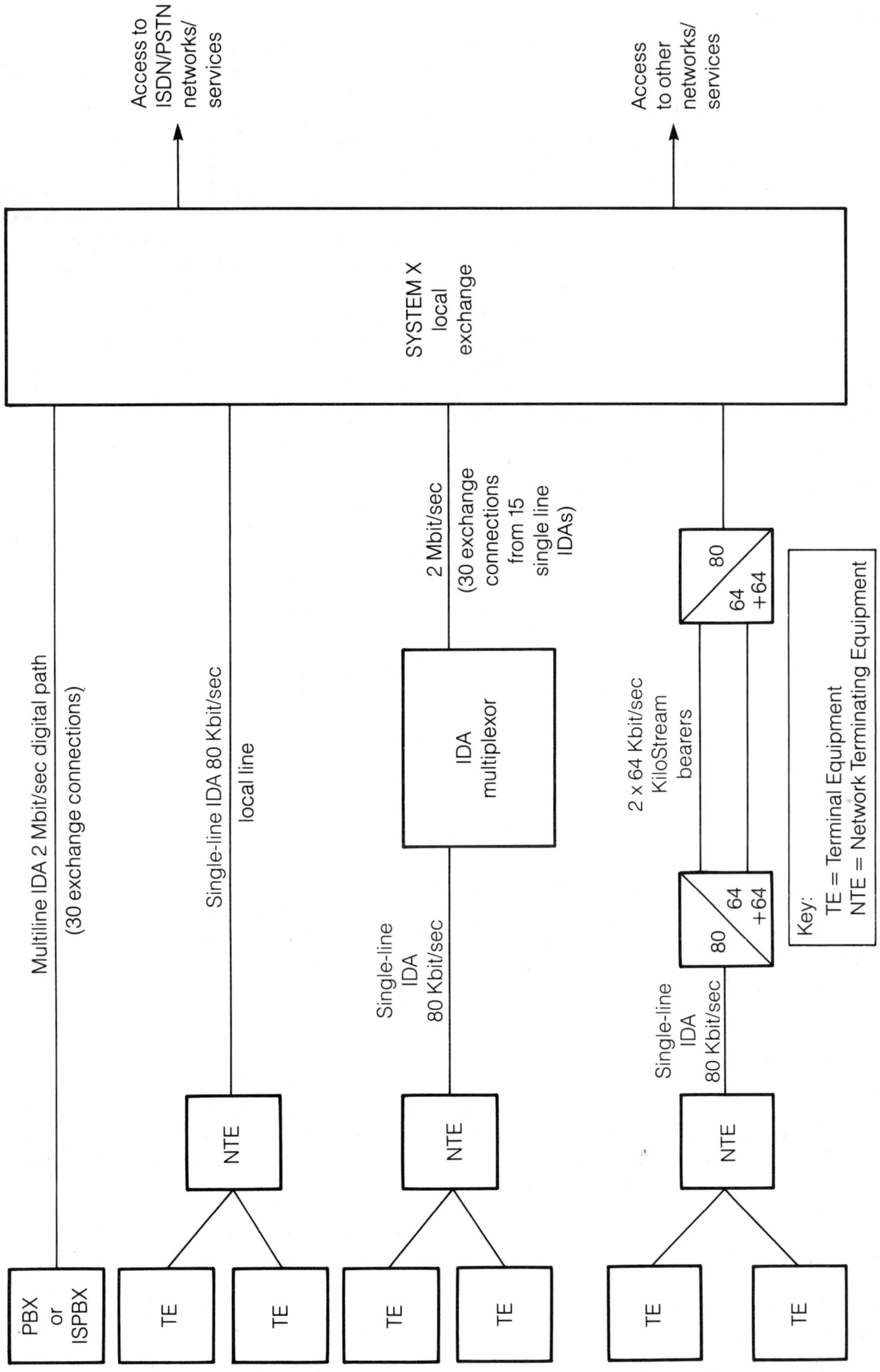

Figure 2: Integrated Digital Access (IDA) implementation and general arrangement of BT pilot ISDN

NTE type	No of digital customer ports	Port types					Channel use B & B′	Remarks
		X.21	X.21 leased line	X.21 bis	Analogue	Modec		
1	1		✓	✓			1. Voice (B) and/or $\leq$8 Kbit/sec (B′) data 2. $\leq$64 Kbit/sec (B) data	Includes integral telephone, autodialler, programmable keys, alphanumeric display
2 (3)	6	✓	✓	✓	✓	✓	Any two ports may use B & B′ as appropriate	Able to utilise existing telephones and modems

Note: NTE 2 development ceased in UK; CEPT NTE2 is equivalent (approximately) to NTE3 in UK

Figure 3: NTE facilities

1 NTE1 (see Figure 4) corresponds to the customer local line termination, as it terminates a single line, and it provides the basic access. It will perform certain functions of layer 1 of the ISO OSI 7-layer model eg line transmission termination, layer 1 line maintenance functions, timing, power feeding and layer 1 multiplexing. In addition it will convert user protocol and user data rates to network protocols eg X.21 to DASS (in UK), protocol conversion and adaptation of the user rate to 64 Kbit/sec. Normally the NTE1 will also incorporate a built-in telephone. A line transmission signal is used to derive 64 Kbit/sec timing supplies in the NTE. This enables the NTE to provide rate adaptation on the data entering at the user's ports — at one of a number of standard rates. The NTE can be programmed to accept data calls at a given rate only, or alternatively at any rate up to a specified maximum. This last feature avoids the need for duplication of ports to allow for the arrival of calls at different data rates. While the digital telephone is in use data services are restricted to the non-voice channel. Call set-up (including on-desk dialling) will be by means of an in-built keypad with an alphanumeric display together with a number of special function and service request keys designed to simplify the use of supplementary services such as Closed User Groups (CUGs) and short code dialling.

2 NTE3, shown in Figure 5, corresponds to user distribution arrangements. It provides the primary access described previously. It is expected that most of their applications will be found in an electronic office, where call set-up is controlled by the user's equipment. For this reason NTE3 has no built-in telephone, keypad, or display; it is, however, capable of accepting a number of user ports which comply with the CCITT Recommendations — X.21 bis, V.24/V.28, X.21 and two-wire analogue.

In some cases NTE3 may play no part at layer 2, and in these circumstances suitable Terminal Equipment (TE) and Terminal Adaptors (TA) will be used.

Local transmission

Local transmission between a given subscriber and the Local Exchange (LE) serving him will be on either a single-line or a multiline basis, dependent upon basic or primary access.

Single-line connection will provide full-duplex transmission over a single pair in the local distribution network. The system will use the so-called WAL2 modulation technique which operates in the burst mode by assembling bits from the primary voice and non-voice channels (the B channels) and the signalling channel (the D channel). The initial stage of ISDN in the UK will use an instantaneous burst rate of 256 Kbit/sec (see Figure 6).

Multiline connection service will be provided on a 2 Mbit/sec digital path. This datastream will be presented to the user at his premises in accordance with CCITT Recommendations G703 and G734.

The line section may use either HDB3 line coding or 4B3T line coding at 1.5 Mbaud.

Multiplexors and local exchange terminations

The programme to replace current analogue exchanges with digital exchanges is likely to be long term. However, there may be potential customers for ISDN in areas not likely to be serviced by a digital LE in the foreseeable future and in such cases IDN multiplexors will be used to provide ISDN services (see Figure 7). Nevertheless, all European PTTs aim eventually to provide digital switching right down to the local level and Figure 8 shows in outline the way in which IDN access will be handled in a digital LE in the UK.

ISDN services

A potential user of any service is interested not only in the method of implementing it; he is likely to be even more interested in the application of that service to his particular business. It is important, therefore, to review the status of the various services that are likely to be provided by ISDN.

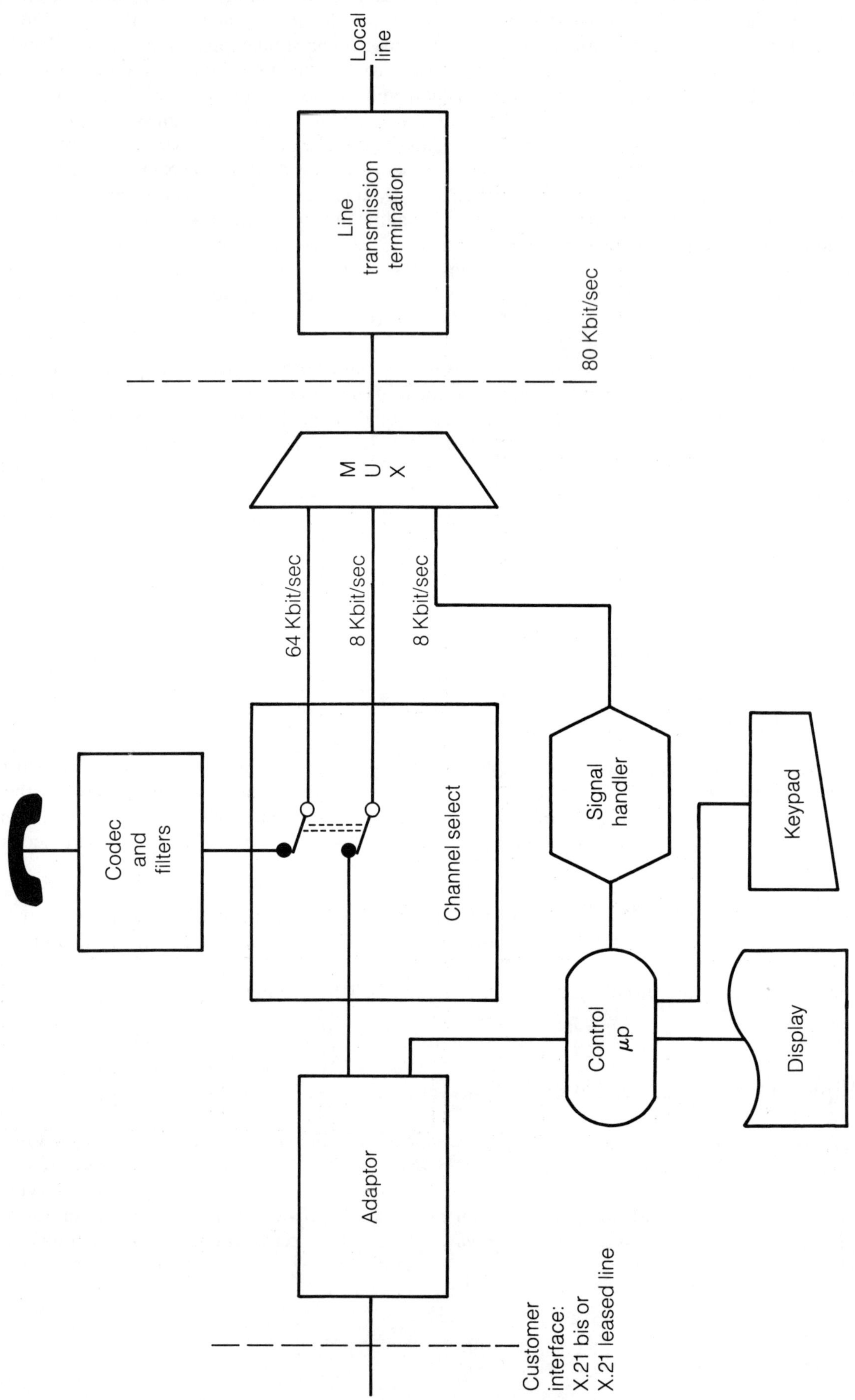

Figure 4: NTE1 functional block diagram

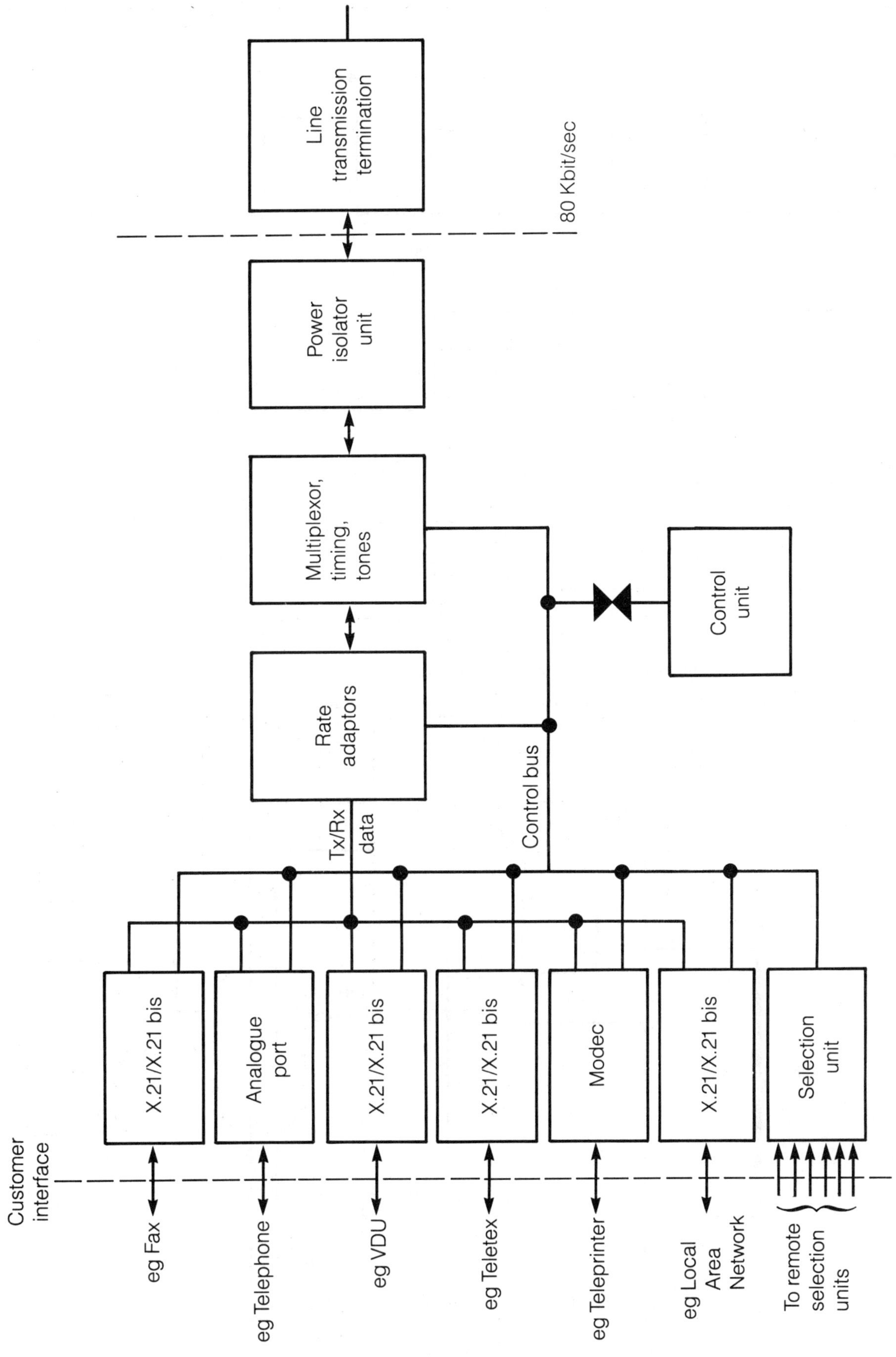

Figure 5: UK NTE3 functional block diagram

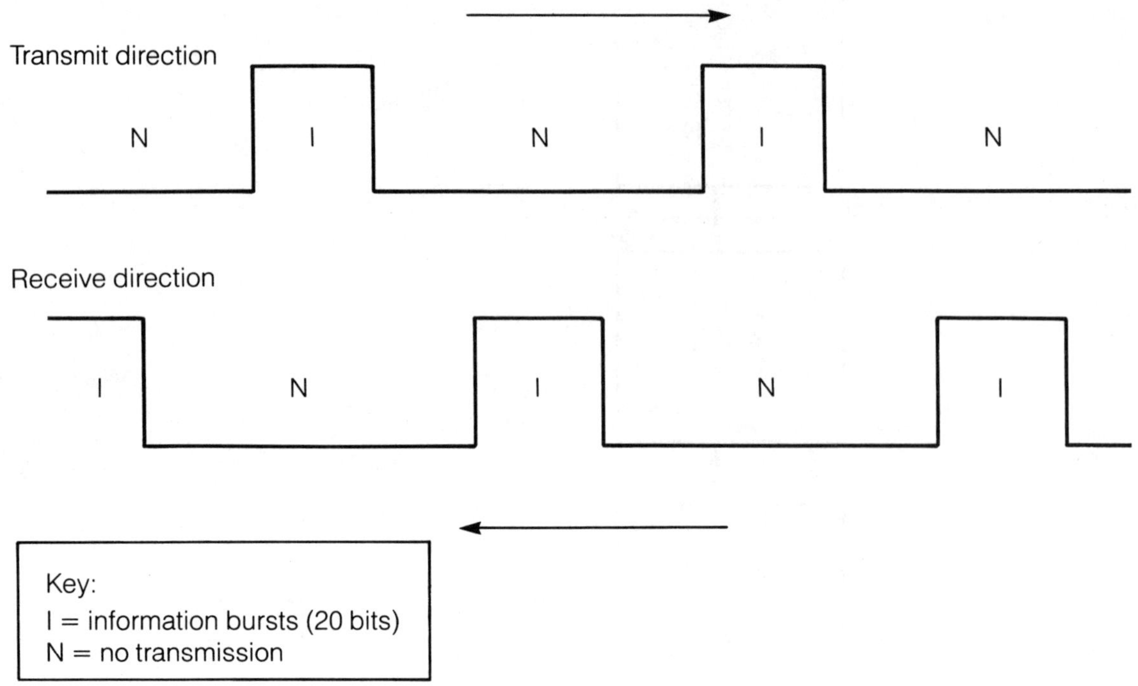

Figure 6: Burst mode transmission (conceptual diagram)

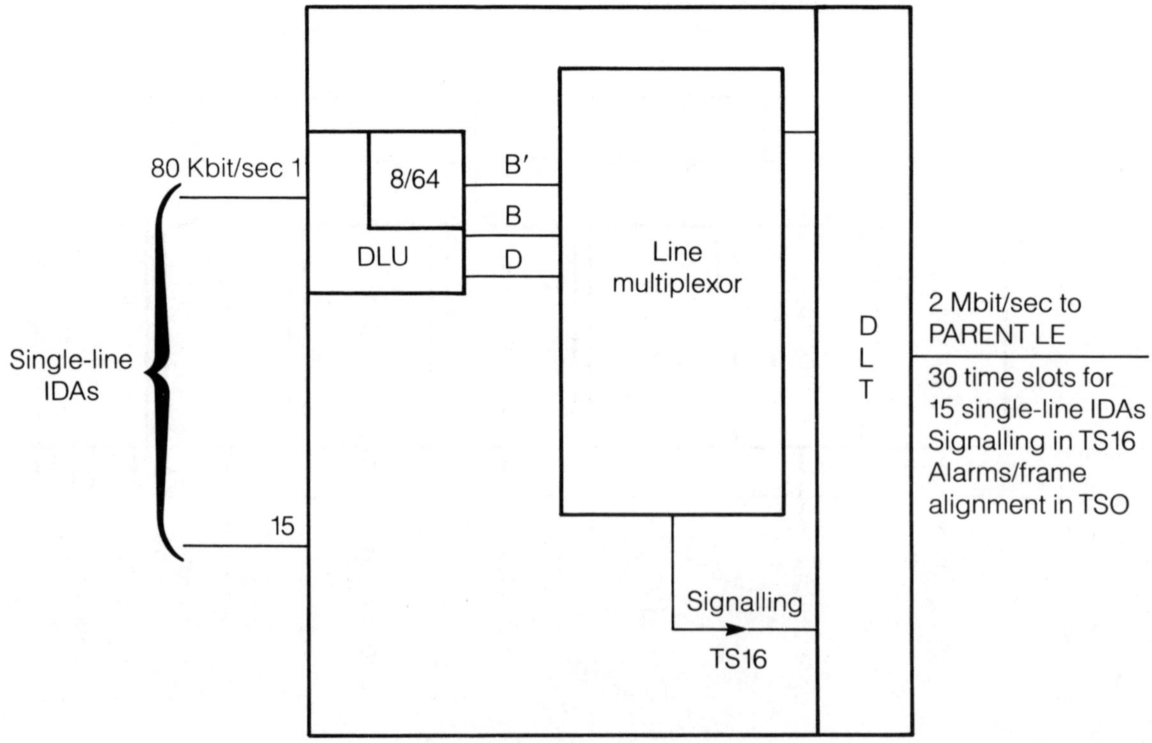

Figure 7: IDA multiplexor

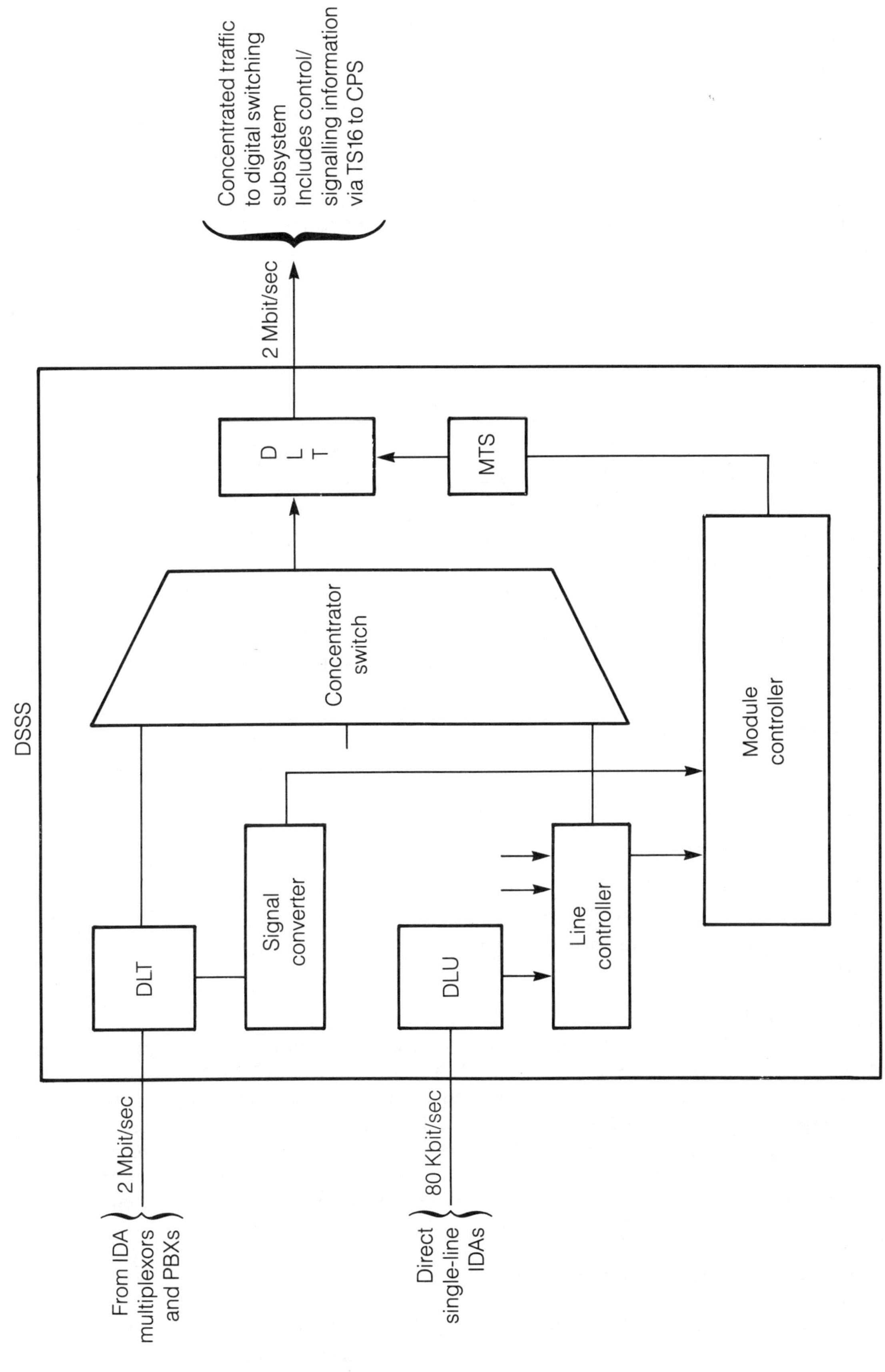

Figure 8: General arrangement of system X digital subscriber switching subsystem for support of IDA

Within the UK, these services may be broadly classified into two categories, as follows:

1 Bearer services which provide the capability for transmitting information between network interfaces and user terminals. The formats and protocols for that information are not defined. This is the so-called 'transparent' service.

2 Standard telecommunication service which provides the complete capability for communication in accordance with defined formats, protocols, procedures and bearer rates.

Both categories will have access to the network-provided supplementary services appropriate to the data and telephony calls. Even on transparent digital connections, ISDN users will be able to make use of Closed User Group (CUG), Originating Line Identify (OLI) and Terminating Line Identity (TLI) supplementary services.

Bearer or transparent services

These services will provide the following.

Circuit-switched synchronous data service

This will operate at user rates of 2.4, 4.8, 8, 9.6, 48 and 64 Kbit/sec, all in accordance with CCITT Recommendation X.1. This service will be offered in conjunction with various user facilities such as CUG, direct call, abbreviated address calling, and line identification. Other facilities may include call diversion, subscriber-controlled barring, multi-address calling and advice of charge. Typical applications for this service include high-speed data transfer (which is so essential for distributed processing), file transfers, high-speed facsimile and slow-scan television.

Circuit-switched asynchronous data service

This service is likely to be provided only on an interim basis. The maximum user rate for this service in the UK is 9.6 Kbit/sec.

Private circuit data service

The ISDN enables either of the user channels to be permanently routed through the network, thus providing a private circuit capability. An example of an application of this service could be for brokers using the private circuit for permanent access to stock market prices, while using the second channel in the switched-mode connection to their customers.

Digital PBX support

If a given digital PBX can support the ISDN access signalling protocols, ISDN will support that PBX. However, user facilities for both telephone and data terminals will naturally depend upon the capabilities of that PBX.

Support for analogue services

Any business introducing new technology and facilities wishes to do this in a smooth planned manner with minimum disruption to current work. This process is sometimes referred to as 'migration'. In a medium to large business organisation, consideration will need to be given not only to the types of application, but also to the geographical locations of branches and subsidiaries. Furthermore, it may be uneconomical to write off all the existing equipment at the same time and replace it by new (even if a PTT should be capable of providing the new facilities simultaneously at all the relevant locations). It is therefore an important facet of ISDN to provide full facilities for analogue ports to which the user's current analogue equipment may be connected eg telephones, analogue PBX or modems.

Standard telecommunications services

The Conference of European Posts and Telecommunications (CEPT) recommends that this service shall be offered as a uniform standard service. It will provide the following facilities.

Access to the packet-switched services

This provides digital access paths to packet-switching nodes for terminals using speeds of 2.4, 4.8, 9.6 and 48 Kbit/sec (ie in accordance with CCITT Recommendation X.1, user classes 8 to 11). Compared with modems and analogue plant, digital access should provide cost and operational advantages, especially at high rates.

An access to Viewtex

This service provides digital access to Viewtex services (in the UK, Prestel) by means of a circuit-switched connection between Viewtex sets and the exchange. This will offer a Viewtex subscriber two important advantages:

1 Access to the Viewtex service without blocking his telephony path.

2 The 64 Kbit/sec connection permits high performance applications such as Videotext (in the UK, Picture Prestel) and high-speed interactive graphics.

A switched Teletex service

This operates at present at the standard international rate of 2.4 Kbit/sec duplex. Low error rates, simultaneous access features and ISDN's fast call set-up times make the Teletex service very attractive to potential users. It will be possible for Teletex protocols to be supported at high rates in addition to the current rate of 2.4 Kbit/sec. Hence it will be possible to use Teletex for both high and low volume applications eg to a data concentrator and a stand-alone terminal. At the high rate (eg 64 Kbit/sec) one page of an A4 document would be transmitted in about 0.5 seconds.

Telephony service

This will enable telephones to communicate with each other using circuit-switched channel connections. The normal range of additional services may also be provided within the ISDN telephony service.

Digitally-switched facsimile

This will offer support to high-definition Group 4 machines with transfer rates for both Group 3 and Group 4. This will be most suitable in electronic office applications as it will be possible to send a document during the course of a telephony conversation while at the same time receiving data by means of the packet-switched service.

Slow-scan television service

This can also be supported by ISDN, providing visual transmission between terminals on either a private circuit or a switched connection. The picture resolution should approach that of the standard commercial television service. The picture refresh time should be about four seconds on the 64 Kbit/sec connection. This will suit applications ranging from surveillance to displaying diagrams or X-ray slides.

A note of forewarning should be introduced at this point. At the time of writing, the CCITT I Series Recommendations defining international interface standards for IDN and ISDN have not been published. This presents a problem to PTTs wishing to offer an early start to the service, for example UK's BT. In order to circumvent this problem, these PTTs propose to offer interfaces to the CCITT Recommendations X.21 and X.21 bis as an interim measure. This approach should insulate the ISDN subscribers from any network changes resulting from the adoption of future international standards or changes in network technology. The range of network terminations based upon the relevant CCITT Recommendations (I-Series) will support the existing range of terminals by the application of a suitable terminal adaptor. The new range of terminals based upon the 'S' interfaces will be supported.

Interworking with other networks

It should be recognised that ISDN will be introduced first in the developed industrial countries, ie into an environment of existing, developed networks. In order to permit access between the different types of users and facilities it is necessary, therefore, to interwork with these existing networks until they are gradually absorbed by ISDN. Consequently a great amount of planning has gone into the question of interworking.

PSTN

Interworking with this network is an inherent feature of ISDN. In addition IDN subscribers with appropriate NTE or ISPBX extensions may interwork the public data-type calls with the telephone network by use of a modem plus analogue port, or a modec.

Packet-switched service

In the UK, ISDN subscribers with asynchronous terminals will be able to access the packet-switched service by means of dial-up connections. Here the calls are set up to the originating exchange by dialling an access number or by using a 'hot line' facility. Once the initial connection has been completed, standard protocols between the ISDN subscriber and the packet-switching exchange-based Packet Assembler/Disassembler (PAD) will identify the parties, thus completing the connection. However an ISDN subscriber's incoming call cannot be supported since the PAD has no means of signalling the destination address to the ISDN.

Subscribers will be able to choose either analogue or digital interworking. With analogue interworking (see Figure 9) subscribers will have to employ interfaces which allow end-to-end compatibility with exchange-based modems. This can be achieved by use of an analogue port plus modem or alternatively a modec port. The modec is restricted to 300 bit/sec, whereas the modem can interwork at any of the dial-up rates. Should digital interworking (see Figure 10) be chosen, shared PAD/NTE combinations are provided in place of the PAD/modem at the exchange. This method relies on a digital path being available across the network to the exchange, but it eliminates the need for modems or modecs.

The synchronous method of access for terminals employing the X.25 interface exploits the capabilities of the ISDN more fully. It offers faster access, eliminating the need for modems. Again two user options are offered:

1 The dial-up service using a Packet Network Adaptor (PNA) (Figure 11) utilises a number of dedicated ISDN ports on the PNA. These are installed principally as a gateway to the Teletex. The PNA is accessed across the ISDN, using an ISDN digital call and the process is completed by the X.25 set-up procedures. The path across the ISDN can be provided by using either the fixed destination call feature or a keypad in conjunction with a remote selector unit, depending upon the type of NTE. Either option supports standard X.25 terminals connected to the X.21 bis interface. Alternatively, should dual X.25/X.25 terminals (eg Teletex) become available, these can employ complete call set-up from the terminal if connected to an X.21 port.

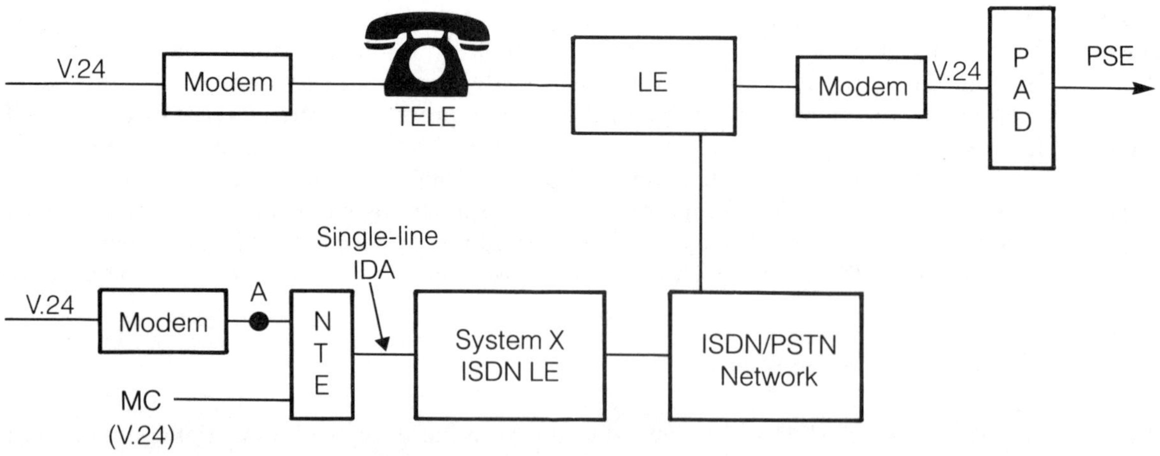

Key:

A	=	Analogue port
MC	=	Modec port
NTE	=	Network Terminating Equipment
PAD	=	Packet Assembler/Disassembler

Figure 9: Asynchronous packet switchstream access: analogue interworking

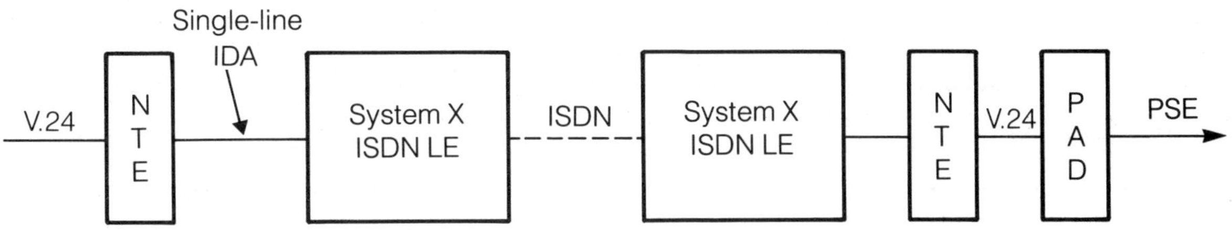

Key:

A = Analogue port
MC = Modec port
NTE = Network Terminating Equipment
PAD = Packet Assembler/Disassembler

Figure 10: Asynchronous packet switchstream access: digital interworking

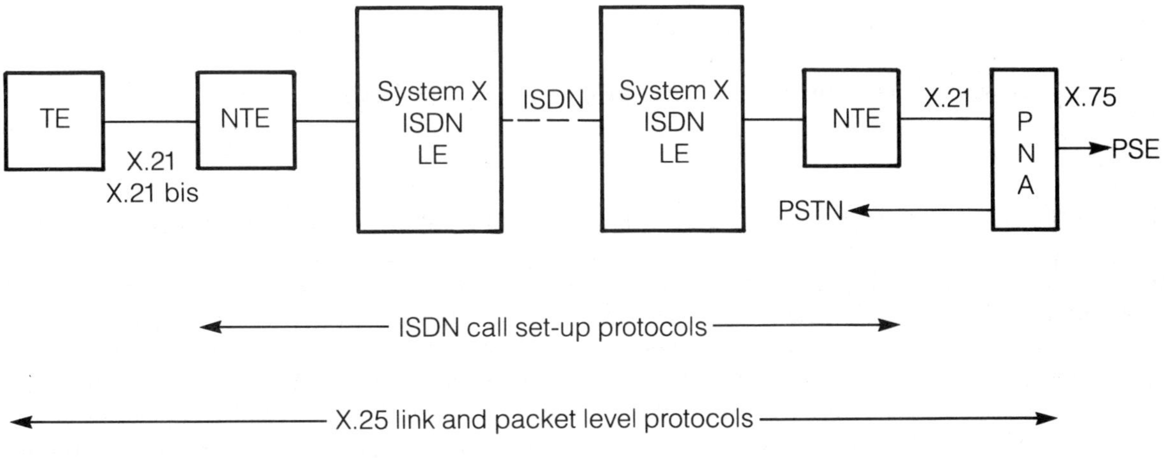

Key:

L = Linker
TE = X.25 dial-up terminal
NTE = Network Terminating Equipment
PNA = Packet Network Adaptor

Figure 11: Synchronous packet switchstream access: via packet network adaptor

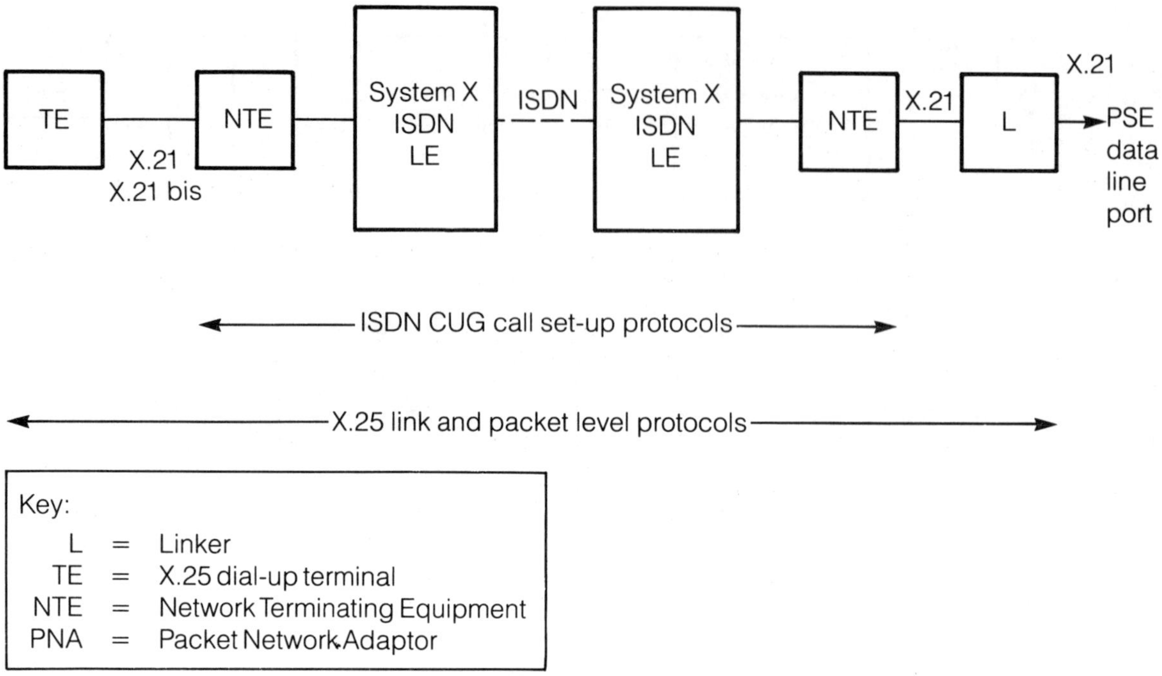

Figure 12: Synchronous packet switchstream access: via dedicated NTE

2 Subscribers who frequently use the packet-switched service (eg ISPBX users with one channel permanently connected to the packet-switched service) may find it advantageous to provide the access shown in Figure 12 as an alternative to using the PNA. This uses an NTE at the exchange which is dedicated to each subscriber. Fraudulent use may be prevented by using either a leased line or the CUG facility on a dial-up access.

International circuit-switched data networks

These are at present being studied but no firm proposals have yet been put forward.

Teletex

Teletex terminals will be supported by the IDN either by use of a modem connected to the analogue port or by means of synchronous digital interfaces. Modem-based terminals will be capable of interworking with each other by using ordinary telephone calls without any interworking units. For interworking with packet terminals the PNA will be needed, while for interworking with telex the Teletex conversion facility as shown in Figure 13 will be required. For terminals based upon the synchronous digital interface, two types of terminal will be supported. The X.21/X.25 terminals will provide call set-up from the terminal in the X.21 mode, which will be followed by data transfer in the X.25 mode. Also, support will be offered for the existing packet-switched service Teletex terminals employing the V.24 interface. Full interworking between IDN terminals of these types will be available across the ISDN using ISDN digital calls. Interworking with all other terminals (including analogue types based on IDN) will be achieved by means of the PNA accessed via the X.21 port, with subsequent interworking using the Teletex conversion facility if the distant terminal is based on the current telex network (see Figure 14).

Interconnection with satellite networks

At present, two implementation phases are envisaged for satellite systems within Europe. The first phase is intended to provide services mainly on a leased line basis by using the European

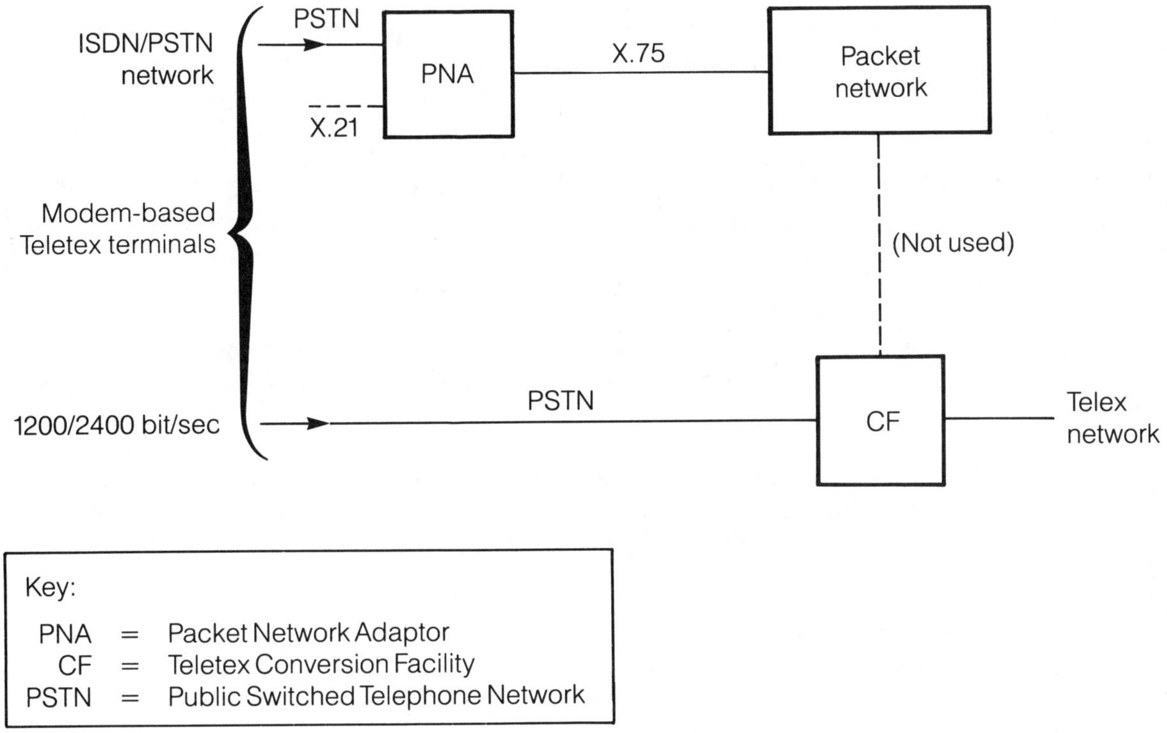

Figure 13: Teletex interworking into PSS/telex networks: modem-based terminals

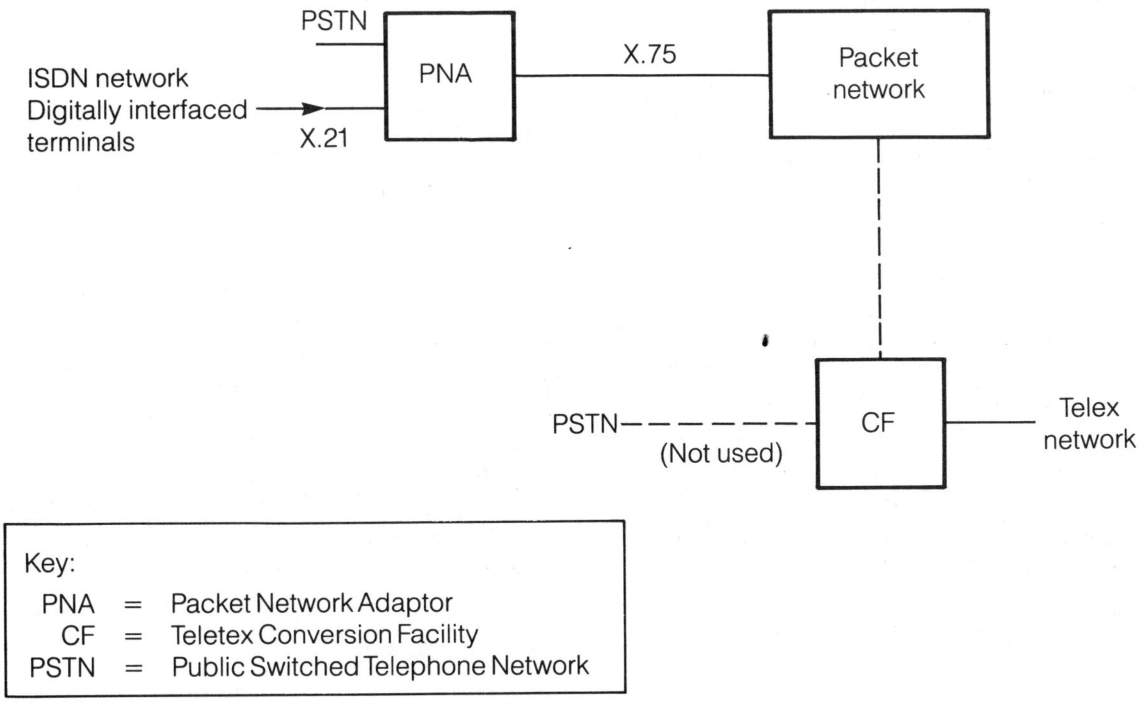

Figure 14: Teletex interworking into PSS/telex networks: digitally interfaced terminals

Communications Satellite (ECS)-SMS which employs the technique of a Telecom I Satellite which employs Time Division Multiple Access (TDMA).

Although the Telecom I-SMS system will largely meet ISDN requirements for switched service, the ECS-SMS system can initially offer only a channel reservation scheme. At some later date adjustment to the subscriber equipment will be necessary to ensure that the interim Intelsat system is consistent with ISDN.

It is expected that, initially at least, large companies will form CUGs using the flexible transmission capabilities offered by satellite systems. Access to a satellite station will consist of dedicated terrestrial leased lines connected to earth stations which could incorporate a switching unit associated with TDMA equipment.

The second phase will be more consistent with the switched services which ISDN offer its subscribers. The demand for audio- and video-teleconferencing services is expected to be significant and the higher bit rates should also be taken into account.

It is expected that the major services provided in the first phase of the satellite systems will migrate to the ISDN as soon as it becomes available. Satellite systems will maintain the function of providing long distance digital paths and they will continue to offer some services requiring different resources such as wideband and broadcast.

Satellites have an inherent wideband capability. The broadcast capability of a satellite system implies an inherent lack of security as the transmission can be easily intercepted. Hence cryptographic units may be required on the satellite links so that end-to-end protection of data can be achieved. In addition, signalling security may need to be taken into account.

A further point that needs consideration is transmission delay. The use of satellites may require some specific routing rules eg to avoid the use of more than one satellite link-routed path in tandem for one communication for voice calls. On the other hand it is possible that non-voice calls, particularly those employing the packet-switching mode of operation, could tolerate longer delays, thus enabling the use, perhaps, of up to three satellite links.

Interconnection with private networks (including LANs)

It must be recognised that for business users integration between communication and office automation and between data, voice, text and image should start right in the 'business area', where Private Automatic Branch Exchanges (PABXs) are generally used. Two kinds of PABX may be identified:

1 The small PABX which is functionally equivalent to a network termination for a basic or standard subscriber access.

2 The large PABX whose functions are much closer to those of a public exchange.

Integrated services PABXs (ISPBXs) are expected to be implemented before ISDN is widely available. They should be able to support the appropriate signalling system such as CCITT Signalling System based on LAP D or, in the UK, DASS. All the circuits of such PABXs could be connected to the local IDN telephone exchange on the basis of 64 Kbit/sec semi-permanent connections. As the network evolves, facilities and combinations of services will be introduced. B-channels between PABXs and the multiservice public exchange will be shared and allocated to a given service on a per-call basis, thus allowing:

- Shared use of resources and better use of transmission facilities
- Access to Value Add Networks (VANs)
- Common procedures for maintenance and operation
- Better adaptation of call routing, depending on the traffic conditions.

The initial functions performed by large PABXs are multiplexing and the distribution of multiservice traffic. Possibly some adaptation functions may also be provided, such as PAD facilities for dataflows generated by asynchronous terminals.

A large PABX should not handle too many signalling systems. Signalling systems based upon CCITT no 7 –

the so-called LAP E (reference: Rec I441 and I450), LAP D (reference: Rec I441) and the D channel protocol, such as DASS, may be considered appropriate. In particular, the message transfer part of CCITT No 7, level 2 and part of level 3, may be considered adequate as a Link Access Procedure (LAP) to support signalling for B channels and to provide transport functions for message-oriented information; this forms the basis of LAP E, mentioned above.

There may also be a need for connection to ISDN for another family of private multiservice equipment — LANs. However, their internal protocols are very specific and hence a gateway will be needed for their interconnection with the public network. Two solutions are at present envisaged:

1 By establishment of a transparent B channel in the public network and communication according to the internal protocol. This mode of operation, however, is possible only if both ends are compatible.

2 By conversion of the internal protocol to the standard protocol used at the public network interface. Such conversion can be simplified if compatibility is achieved at the LAN definition stage.

Future ISDN developments

Even as the initial stages and pilot schemes regarding ISDN are being implemented, ISDN concepts are continuing to evolve and further stages of development are being considered.

As the ISDN evolves, there will be a trend towards accepting the CCITT I Series Recommendations as a universal access (see Figure 15), thus resulting in the declining use of the V and X series interfaces. Some of the aspects currently being specified include:

- Improved forms of local two-wire transmission
- NTEs with two X.21 ports and adaptors to support other services
- A more flexible structure allowing subscribers to use one of the B channels for either the circuit- or packet-switched mode of data transport
- A wider range of interworking capabilities
- Enhancements to the call-processing and signalling subsystems, to increase non-voice service capabilities in terms of user-to-user signalling, network address extension for subaddressing of terminals and support for the full capabilities of intercommunicating ISPBXs.

Other aspects are also receiving full consideration. These include:

- Improved maintenance facilities for ISPBXs where messages will be conveyed in time slot 16
- Configuring the 64 Kbit/sec B channel as a larger number of lower bit rate channels eg to maximise the use of speech compression techniques.

As the evolution of ISDN is being considered, it is also important to consider possible future developments of the supplementary network capabilities.

Although storage capacity and processing power can be spread throughout the ISDN to meet most user and administrative service requirements, some services require so much extra storage and processing that they are best located at a few 'special' nodes in the network or, alternatively, provided as a VAN. The sophisticated signalling capabilities of ISDN can be exploited to control such network-provided services. The special node and VAN approach may be an appropriate method of providing such capabilities and ISDN should be capable of interworking with them.

A number of examples may be considered in this context. These include:

- Voice message storage for network-provided messages and answering services and user services such as voice store-and-forward
- Data message storage and retrieval features to satisfy the needs of various data and text services, including store-and-forward, delayed delivery, mailbox, broadcast and multi-address message services
- Combinations of voice and data storage and retrieval features to support data-in, voice-out services and (when the technology permits) voice-in and data-out services
- High-capacity storage services, such as Viewtex, where the information supplied by other providers is being accessed (including interworking by means of protocol conversion in cases where the subscriber has a printer or a visual display terminal)
- Means of charging in a variable manner for the services provided at the special nodes.

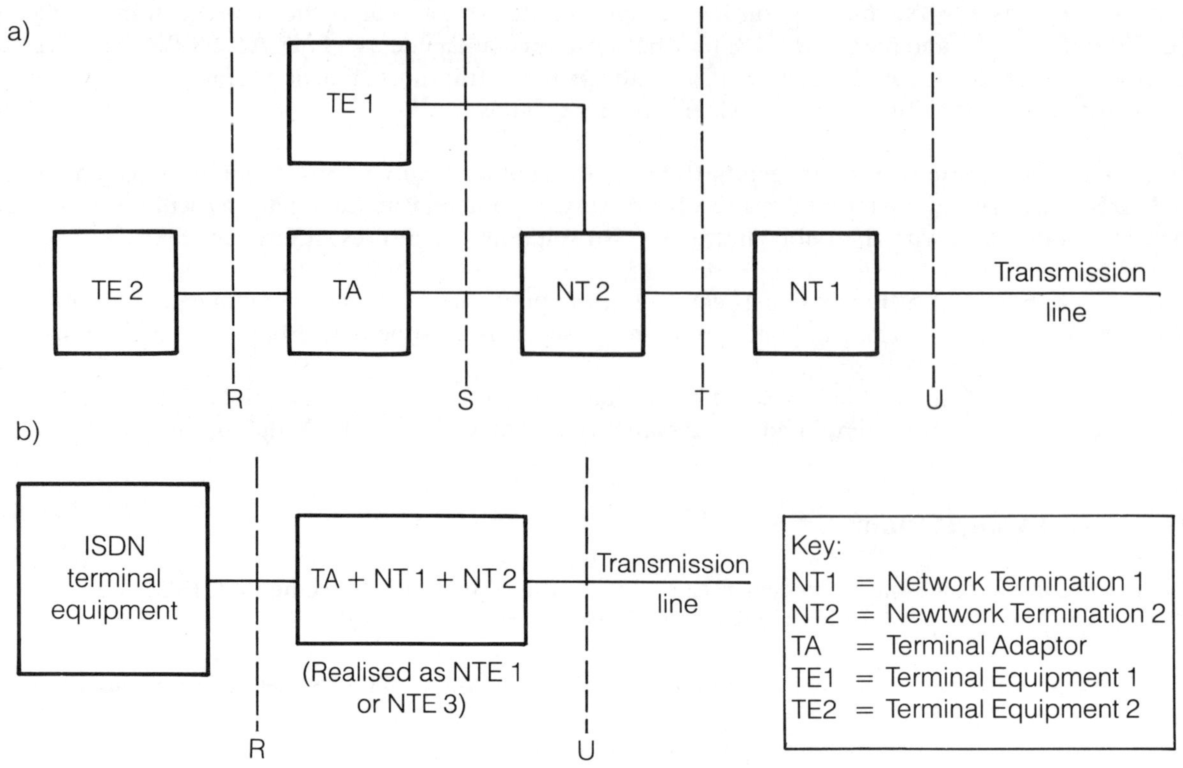

Figure 15a: CCITT generalised I-Series reference configuration
Figure 15b: Variation of reference configuration for BT pilot ISDN
Figure 15: CCITT I-Series Recommendations

The current international discussions at CCITT are aimed at specifying a uniform series of recommendations (the I Series) covering all aspects of ISDN. The recommendations dealing with ISDN user/network interfaces are of particular interest. These cover reference configurations, channel structures and access capabilities. The eventual ISDN user/network interface is likely to be based upon 144 Kbit/sec, consisting of two 64 Kbit/sec circuit-switched channels and a 16 Kbit/sec message-structured channel for signalling and packet-switched data.

Subject to the normal commercial and regulatory considerations, both national and international, most PTTs expect to implement the CCITT ISDN standards progressively once they have become firmly established. However they propose to continue to support any interim standards that have been nationally adopted, thus dispelling users' fears of potential obsolescence of their equipment and overcoming their reluctance to make an early commitment to the service.

Acknowledgement

The author wishes to thank British Telecom for information supplied on ISDN and IDA.

6: Data security in Open Systems Interconnection (OSI)

W L Price

National Physical Laboratory
Teddington
UK

This paper examines the way in which standards for data security are being developed, paying particular attention to that part of the work relating to Open Systems Interconnection (OSI). As yet, these developments are incomplete, but considerable progress has been made.

W L Price

Mr Price, a Principal Scientific Officer at the National Physical Laboratory, Teddington, is responsible for the Laboratory's work on data security applications. Prior to this he was in charge of the research into the behaviour of computer networks using simulation techniques. He is a Fellow of the Institution of Electrical Engineers, a Fellow of the British Computer Society and a Senior Member of the Institute of Electrical and Electronics Engineers. He is co-author with Donald Davies of the forthcoming book 'Security for Computer Networks' (John Wiley and Sons).

Data security in Open Systems Interconnection (OSI)

Introduction

Now that Open Systems Interconnection (OSI) has reached the stage where the architectural model *(PRI1)* is agreed upon, attention is concentrated on the development of communication protocols that fit within the architecture at the various levels. Among the services that may be included in these protocols are those designed to provide security for the data handled. This paper surveys the ongoing work to establish the data security enhancements of communication protocols.

First we must define what is meant by data security, which may mean different things in different contexts and to different people. An obvious requirement is for privacy, where intruders are prevented from reading data not intended for their eyes; this can be very important where data links are susceptible to tapping or data may be misdirected to the wrong destination within a network. We shall classify techniques designed to prevent disclosure of data under the heading 'secrecy'. Equally important is the need to prevent the undetected alteration of data that is being handled in a communication system. This property has been called 'authentication' of data; the term is used in this sense in some US national standards documents emanating from the banking community. Unfortunately, the term is also used within the OSI community in the sense of peer entity authentication, where communicating parties seek mutual verification of each other's identities; this is another important requirement in the communication security area. For this reason the term 'integrity' of data is now being adopted fairly widely to describe the process of preventing undetected alteration, reserving 'authentication' to mean peer entity verification. Complete prevention of alteration is frequently impossible; what is possible is to detect alteration when it happens.

A fundamental technique in achieving secrecy, integrity and authentication is that of encipherment, where data is transformed by a complex algorithm under the control of an encipherment key. Within the last 10 years considerable progress has been made in the development of encipherment algorithms for use in the public domain ('public' as distinct from the world of military and diplomatic data protection). Publication of the US Data Encryption Standard (DES) *(PRI2)* has resulted in its subsequent adoption as an American national standard (known as the Data Encryption Algorithm (DEA) *(PRI3)*; it is also being considered as a possible international standard. Responsibility for US federal DP standards lies with the National Bureau of Standards of the Department of Commerce; US federal telecommunication standards come under the National Communications System. Responsibility for US national standards lies with the American National Standards Institute (ANSI). Each of these bodies has been involved in the preparation of standards for data security.

Outside the US, the British Standards Institution (BSI) has played a leading role in the preparation of standards for data security; in this work the responsible committee is OIS/21. On the international scene the International Standards Organisation (ISO) set up Working Group 1 of Technical Committee 97 in 1980 to consider the need for standards for data security; this working group has recently been given the status of a full subcommittee (SC20) of TC97 and held its first meeting in Birlinghoven, West Germany, in January/February 1984.

Existing standards for data security

The DES (and DEA) defines a block cipher algorithm with 64-bit plaintext input, 64-bit ciphertext output, under the control of a key of 56 bits (64 bits including parity). In addition to the standard definition of the algorithm it has been necessary to define modes of operation for the algorithm. The standard modes are Electronic Codebook (ECB), Cipher Block Chaining (CBC), Cipher Feedback (CFB) and Output Feedback (OFB). US federal (FIPS 81) *(PRI4)* and national *(PRI5)* standards exist to describe these modes. They are concerned with the way in which the algorithm is used to encipher data bits, bytes or blocks; they do not describe how encipherment may be fitted into a communications context. Use of the DES to provide data integrity (or data authentication) has been defined in an appendix to FIPS 81 and in an ANSI standard. This is effectively another mode of operation and is gaining wide adoption within the banking community both in America and Europe.

In passing, we can observe that products embodying the DES are available from a wide range of sources, mainly in the US. The algorithm block execution time in these products ranges from 100 msec (microprocessor) to about 4 μsec (special-purpose LSI).

Since the DES emerged before the concept of OSI it is not surprising to find that efforts to incorporate encipherment in communication protocols began at a time when some lack of order existed in the world of protocol definition. This work began under the auspices of the US National Communications System, which published a series of drafts of a federal standard which initially went under the name 'Telecommunications: Interoperability and security requirements for use of the Data Encryption Standard in data communication systems'. As the OSI work developed, so the name of this draft standard changed. Initially, the draft aimed to define security enhancement at what became OSI levels 1 to 4 (and, to some extent, even above level 4). However, it was seen as being impossibly ambitious; gradually the scope was narrowed down, so that the only standard publication, FS 1026, that has so far emerged from this work is now entitled 'Telecommunications: Interoperability and security requirements for use of the Data Encryption Standard in the physical layer of data communications' *(PRI6)*.

FS 1026 is a very concise document, relying on FIPS 46 and FIPS 81 for basic definitions of the algorithm and modes of use. Apart from providing for interoperability of equipment, the sole aim of this standard is to prevent disclosure of plaintext; in other words, data secrecy is provided, but not integrity. The specified mode of operation is 1-bit cipher feedback; this can be applied to both synchronous and asynchronous communication. Under asynchronous operation two methods of dealing with break conditions are specified as alternatives; also it is possible, as an option, to restore parity on encrypted characters. Encipherment keys are either to be loaded into the device from a manual keyboard or from a key-carrier module interfacing with the encipherment unit.

An ANSI standard was published almost simultaneously with FS 1026; this is entitled 'American National Standard for Information Systems — Data Link Encryption', X3.105-1983 *(PRI7)*. Despite its title this standard covers encipherment at both physical and link levels of OSI. The purpose of the standard is 'to provide interoperability between Data Encryption Equipment (DEE) within a link data communications environment'; its main aim of interoperability is therefore similar to that of FS 1026, though the latter confines itself to the Physical Layer only. However, in the scope clause of the ANSI text one finds the statement that 'it is recognised that several cryptographic devices for Data Encryption Algorithm (DEA) that predate the publication of this standard exist. Where the operation of this equipment conflicts with the preferred method of operation described, their method of operation is identified as an acceptable alternative'. Since interoperability is a declared central aim of the standard, this is a rather strange statement, since it implies that equipment conforming to the standard may, because of the predating clause, not be interoperable. Furthermore, the standard contains a number of alternatives in the detailed definitions of the protocol enhancements. There is some confusion between the terms 'alternative' and 'option'. In this paper we shall assume that 'alternative' means a different way of carrying out an essential function, common to all DEEs, while 'option' means an additional function which is not necessarily present in every example of a DEE; this is the usage in FS 1026 and seems to be well-chosen terminology. In the ANSI standard we have the terms 'preferred options' and 'options', meaning respectively preferred alternatives and other alternatives — not preferred.

It is beyond the scope of this paper to list all the alternatives and options contained within X3.105, but we must mention the significant ones. The most important alternative relates to operation at the link level; here we have two methods of encipherment — I-field encipherment (the preferred method) and full-frame encipherment (the alternative method). In I-field encipherment it is the data link service data unit which is enciphered, as received from the Network Layer above. In full-frame encipherment it is the data link service data unit plus the protocol control information of the Data Link Layer which is enciphered; this is almost synonymous with the physical service data unit. In topological terms the former function is located near the top of the Data Link Layer, while the latter function is located near the bottom of the Data Link Layer. The effect of full-frame encipherment is to limit the entities that can make use of the protocol control information to those that actually possess the encipherment key (it is assumed that they are capable of carrying out the encipherment algorithm); this could mean that in a multi-drop context all stations sharing the line would have to use the same key. This disadvantage is not a feature of I-field encipherment. We shall return to the question of the location of the encipherment function within an OSI layer when we consider the broader aspects of security in OSI.

At both Physical and Data Link Layers, the recommended mode of operation in X3.105 is cipher feedback; at the Physical Layer the size of the feedback path is one bit, while at the Data Link Layer the size of the feedback path is one character wide for character-oriented encipherment and eight bits wide for bit-oriented encipherment. A wide range of lengths of initialising variable is specified, with a preferred alternative for bit-oriented operation and for each size of character in character-oriented operation. An option at both layers is bypass control; this enables a DEE to be bypassed (with appropriate local or remote control commands) for circuit-testing purposes. The existence of a bypass in a DEE is regarded by some as an undesirable feature, possibly exploitable by an intruder in causing plaintext to be transmitted unintentionally.

Evidently, a purchaser of equipment claiming conformance with X3.105 must enquire carefully which alternatives are supported; otherwise he may find that new equipment may not be able to communicate with equipment purchased earlier even though it conforms with the same standard. FS 1026 at the physical level allows no serious incompatibilities.

Security in OSI

The requirement for security in OSI was recognised early in the preparation of drafts for the standard that has eventually emerged. Encipherment capability was envisaged at three levels — network, transport and presentation — in the draft versions. In the final standard encipherment is only mentioned at the presentation level. The philosophy behind this seems to regard encipherment as a kind of code transformation (which in some ways it is) and therefore an appropriate function to place in the Presentation Layer, where transformations of representational code are an important part of the role of the layer, providing compatibility between diverse users. On the other hand, this statement must not be taken to indicate that the presentation layer is the only appropriate location for encipherment within OSI. Since each layer in OSI adds its own header information to the data service unit received from the layer above, if concealment of header information for data in transit is required, encipherment must take place low in the protocol hierarchy. Reading of headers from communication links may yield information of value to intruders, giving the source and destination of messages and the number of messages passing between particular pairs of users, even though the actual data itself may be enciphered and is therefore unintelligible. However, if encipherment takes place low in the hierarchy, then it is remote from the user and therefore not directly under user control; furthermore, the traffic of many different users may be carried under the protection of encipherment with the same key, which may not be acceptable for sensitive information.

Another issue affecting choice of encipherment function location is the extent of the encipherment domain; encipherment at the session layer or above in OSI means that the data remains in enciphered form from one end of the network (or networks) to the other. On the other hand, encipherment at a low layer, for example the Data Link Layer, means that data is protected in transit between neighbouring network nodes, but becomes plaintext within each node traversed. The latter is obviously not acceptable for any data of any appreciable degree of sensitivity.

There is a case for providing each kind of protection — user-specific encipherment and concealment of headers, end-to-end and node-to-node encipherment; this can be achieved by installing encipherment at two hierarchical levels — one low and one high. On the other hand, since encipherment usually involves some overheads, be they extra cost of equipment, complication of protocols, data expansion or bandwidth limitation, it should not be installed at more levels than necessary. In the American work referred to above, there has been a tendency to begin at the bottom and work upwards, specifying how to install encipherment at each level. Since this work has not progressed very far in definitive form, it remains to be seen at how many levels encipherment is finally specified by the federal authorities or ANSI.

There has been considerable discussion on the question of which committees shall be responsible for specifying security enhancement of communication protocols; generally speaking it seems that security definitions may be contained in addenda to communication protocol standards. It seems likely that data security will come to be regarded as a 'quality of service', in much the same way as error correction and detection and other qualities of service. The main bulk of the work on secure protocols has been done so far by the committees directly concerned with data security and not by those whose responsibility covers the communication protocols themselves, though there are strong signs that the latter are becoming aware of the need. There is a danger that the committees on security will consider that the communication committees know little about security; conversely, the communication committees may think that the security committees know little about data communications. In practice this is unlikely to be true, although, since the preparation of standards is a complex matter, with much give and take on many issues, some of the decisions taken may not seem logical to those who have not taken part in the work. In the UK the BSI committee responsible for data security, committee OIS/21, began its work in 1980. The committee responsible for communication protocols, OIS/16, set up a working group, WG13, in 1983, to consider security enhancements to communication protocols. Fortunately, there is considerable overlap in the membership of OIS/21 and OIS/16/WG13, so that disharmony in the work is being avoided.

One decision that must be taken is the identification of the OSI layers in which encipherment is permitted. There has been some argument that encipherment ought to be allowed at any layer that an implementor may choose, but this could lead to widespread incompatibility of implementations. There is, therefore, a case for choosing a limited number of layers and requiring that, for conformance to OSI standards, encipherment shall only take place at these layers. It is too early to try to identify in this paper which layers will be chosen, but we may observe that there is some opposition already to choosing the Data Link Layer. As we have seen there is already an ANSI standard applicable to this layer. The rationale for the opposition is based on the premise that Data Link Layer encipherment confers little that is not obtainable by Physical Layer encipherment; if the integrity of data is to be protected, as well as secrecy, then this could better be provided, it is suggested, by a security enhancement at a level higher than the data link.

When the candidate layers for data security enhancements to protocols have been selected, there remains the issue of the location of the encipherment function within the layers themselves. We have seen that ANSI standard X3.105 already contains alternative locations within the Data Link Layer. In the broader context the number of alternatives should be strictly minimised for reasons of interoperability. There is much to be said for the encipherment of data service units as they are accepted from the layer above; this would be a 'clean' solution which could be generally applicable throughout the hierarchy. However, this may be too simplistic a view, because some functions, authentication of data for example, may interact closely with the protocol of a layer. If a message fails to authenticate, then the layer may wish to request a retransmission, rather than pass a faulty message to the layer above. A security feature of this kind might be best located somewhere within the body of the layer, rather than at the top edge. Thus we see that the choice of location may not be straightforward.

An essential element in any data security structure is the provision of keys for encipherment and authentication. At the physical level there is little choice in the manner of key management — keys must be loaded by human intervention; this can be done on a keyboard or via a key transport module. At the higher levels there is the possibility of transporting keys via the communication channels that are being protected; for many purposes in a widespread network or chain of networks this is the only practicable solution. For this purpose, it is possible to set up a hierarchy of keys, data encipherment keys and key encipherment keys, the latter being used solely to protect data encipherment keys from disclosure while in storage or transit. Communicating entities wishing to set up a secure connection may need first to be in possession of mutual key encipherment keys. The actual source of data encipherment keys may be the calling entity in a liaison or there may be a special server attached to the network whose function is to provide keys for secure connections on demand. The key server scenario has the advantage that key encipherment keys need not be in existence between every potential pair of communicating entities. A discussion of key management methods may be found in *(PRI8)*. It goes without saying that choice of keys must be such that an intruder cannot predict or discover what keys are being used.

In the work of preparing communication protocol standards for use in OSI, attention is now turning to the need for management protocols which will be concerned with controlling the protocol hierarchy. Clearly there must be data security features within management protocols. It is only by some such control system that application processes can request that lower layers provide the required data security facilities and, indeed, check that these facilities are present and operating correctly. 'Vertical' control channels need to be provided for this purpose. It is very likely that provision of encipherment keys will need to make use of these vertical channels.

Conclusion

Because the development of secure communication protocol standards is still in progress, this paper must necessarily end on an incomplete note. However, work on standard communication protocols within OSI is rapidly gathering momentum. The need for data security enhancements for these protocols is being recognised and work is in hand to produce the necessary definitions. The full structure is as yet far from complete and much remains to be done.

7: Planning for telecommunications

J Shields

Standard Telephones and Cables plc
London
UK

The use of computers and man/machine interactions is propelling information technology into an increasing role in the social, economic and cultural affairs of our daily lives. The accelerating need for easy access to information services is generating the free movement of larger volumes of information. A suitable telecommunications infrastructure is thus essential to these developments.

J Shields

Dr Shields' career over the past 25 years has been in industrial science and engineering, and in 1981 he was appointed full-time Managing Director of STL, the main research and development arm of Standard Telephones and Cables in the UK. Over the previous five years he was Deputy General Technical Director of the world-wide operations of ITT and Vice President of the ITT system in Europe, with responsibility for the activities of 15 000 scientists and engineers throughout all European companies of ITT, engaged in the telecommunications, semiconductor and components industries. Between 1971 and 1976 he was responsible for the business management of the commercial and industrial products of ITT in Europe, prior to which he was Managing Director of GEC's semiconductor operations in the UK, having been involved in the technical development of semiconductors since the 1950s.

Planning for telecommunications

Introduction

Since people first learnt to communicate through the power of speech, the need to communicate over ever-increasing distances has grown. Audible methods, such as shouting and the talking drums of Africa, have a very limited range. However, Man soon realised that light could be seen from afar. Indeed, the fall of Troy was signalled from Argus to Mount Ida by a chain of beacon fires. Such beacons are an example of a simple binary code. The light from the fire indicates that an expected event has taken place, not even that the danger of the event taking place has passed.

Bit rates and information content can be considerably increased by the use of devices employing codes such as semaphore, the heliograph and the Aldiss Lamp. Even with these, speed is limited by the muscular reactions of the operators to a few bits per second.

It was not until the invention of the telephone in 1876 that a few thousand hertz could be transmitted over relatively long distances.

We now have systems with bandwidths of hundreds of megahertz and digit rates of hundreds of megabits per second. Information technology is driving transmission rates even higher.

Information transfer

The use of computers and man/machine interactions is propelling information technology (IT) into an increasing role in the social, economic and cultural affairs of our daily lives. The accelerating need for easy access to information services is generating the free movement of larger volumes of information. A suitable telecommunication infrastructure is thus essential to these developments. Two computers in adjacent rooms can pass information between themselves at increasingly higher rates. The objective of telecommunications is to remove the problems of distance so that computers on opposite sides of the world can converse as easily as those in adjacent rooms.

Until the early 1970s our telephone systems were analogue. This meant that in order to pass data we had to convert the digits into tones by means of modems. The rate at which this could be done was limited by the bandwidth of the telephone system to a maximum of about 2.4 Kbit/sec. Then in the 1960s and 1970s digital transmission was introduced into the telephone network. Digital switching is now beginning to follow and is being merged into what is known as the Integrated Digital Network (IDN) in which connections established by digital exchanges will be used for the transmission of digital signals for a single service such as telephony.

Telecommunication services

As the need for other services such as videotext and data transmission increases, services to provide these can be separately provided and expanded. The alternative is to have a single network that would evolve from the existing telephone network. At present there are separate networks already existing and developing, such as the following:

- Telephony
- Telex
- Facsimile
- Teletext
- Mercury
- Cable TV
- Satellite TV
- Private networks (including data).

(In addition, if more wideband circuits could be made available to the end users, particularly the domestic subscribers, then, although initially used for only one-way communication, the potential for two-way use opens up a wide range of new services from the videophone through electronic mail to banking and shopping from the home.)

The use of one network employing the same digital exchange and digital paths to establish connections for all the various services has been named the Integrated Services Digital Network (ISDN). Multiservice networks (such as ISDN) have a number of advantages. It is easier for the administration to manage a single network than many with different characteristics. Capital is better utilised and planning is easier, especially for new services where there is a lack of information on future likely demand. The same type of equipment may be used and ordered in large quantities, and new facilities can be introduced more easily in the entire network. Operational staff can be trained more easily and used more efficiently.

For the user, and especially for the small user, there are operational and cost advantages since only a single access is required to a common network rather than separate telephone, telex and data lines. The usefulness of a service grows as the number of potential subscribers increases, and the ISDN provides the maximum connectivity for all users. However, this is only likely to happen if the ISDN can provide a satisfactory level of performance for data users at an economic cost. Performance aspects need to be studied in conjunction with the application of non-voice services, since the cost of a network is closely related to its performance. The cost of telephones, the basic service, should not increase with the introduction of the ISDN.

New terminals are being developed for the range of telecommunication services now being introduced, such as teletext, videotext and facsimile. Initially, these services will be provided on separate terminals, but integrated workstations will gradually be introduced, providing a wide range of facilities and services from a single compact unit. However, use of the new integrated services will lag behind their introduction because terminal equipment will most probably be paid for by users rather than the administration. Users will not invest heavily in specialised terminals until they can see that they will be widely used as well as being cost effective relative to existing services.

Integrated services network

Non-voice (eg telex and data) telecommunication at present represents a relatively small number of subscribers when compared to telephony. Consequently, small amounts of capital are required but at a high cost per connection. As these services grow without the advent of ISDN, new investment becomes comparable with the existing investment in the telephone network. However, if the growth of separate networks is slowed and new non-voice subscribers are connected to ISDN, while it cannot be predicted that no additional cost will be incurred for ordinary telephone subscribers, there is a condition for the economic acceptance of the ISDN concept that this cost should be negligible.

How can such a single multiservice network evolve from the apparent increasing number of individual systems? Only by some form of international standardisation. The International Standards Organisation (ISO) has developed an 'Intercept Strategy for Open Systems Interconnection'. An 'Intercept' is a recommendation rather than a standard but can be introduced much more quickly than an international standard.

The ISO Intercept has identified a Reference Model with seven layers. Each layer is self-contained and only the interfaces to adjacent layers, and the service it provides, need to be fixed. Each layer uses services provided by lower layers in order to carry out its functions. Functions within a layer but in separate systems communicate with each other using protocols appropriate to that layer. No protocol acts across layer boundaries. Briefly, the layers are as shown in Figure 1. Such standardisation provides the only way to the exploitation of a successful ISDN.

New technologies

The two most significant technologies that will advance telecommunications into the 1990s are optical technology and processor technology and software.

Optical technology

Optical technology was first envisaged at STL in the UK in 1966. Since then this concept has been developed and its impact on transmission and central transmission is seen to be very significant. As it offers virtually infinite bandwidth, it is a technology that can free us from the dilemma of bandwidth restrictions. This is the basis of research now being undertaken. New materials for lasers, detectors and fibres extending their usability to wavelengths beyond 1.3μ and 1.5μ are significantly reducing losses due to attenuation and dispersion.

At present, most processing of the signal (eg switching and amplification) is performed electronically with conversion being required between optical and electrical signals. Perhaps the greatest opportunity in optical technology lies in this area. Optical transmission is presently at a similar stage to the time in radio development of the spark gap transmitter used for the modulation of radio signals. The challenge is to control the detailed level of coherency of the basic signal and to switch and regenerate that signal in the optical mode. This will enable phase and other 'secondary' aspects of the signal to be used for the transfer of information so that, for example, bandwidths of 10 000 GHz become feasible. This promises to enlarge the information handling capability of the optical signal.

Optical communications will certainly be the technique used for all new systems; it will also be used to

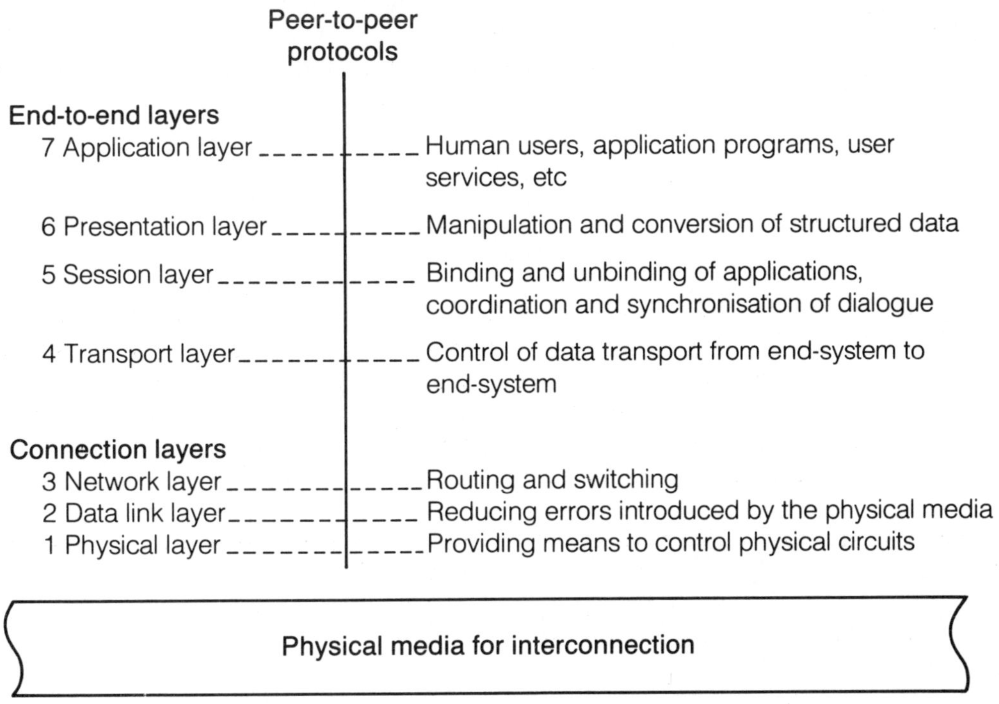

Figure 1: The ISO seven-layer reference model

extend the capabilities of our current installed transmission systems.

Processor technology and software

By 1990, fail-safe, multiprocessing, high availability micro-mainframe computers will be available for incorporation into our telecommunication systems. In parallel with these processors will come the software design and implementation languages. Clearly we cannot, nor should we try to, envisage this as independent development. Both will come together. Complex software systems will be achieved in the same way as hardware systems; ie since we have been able to modularise and interconnect modules of hardware, we shall similarly be able to tackle software systems. Until we can modularise or define interfaces at the subsystem level, we shall not effectively be able to produce large complex software systems.

Conclusions

What does all this mean to the DP manager? In the early days of commercial computers, and it is difficult to realise that this was not much more than 25 years ago, batch processing was the norm. DP managers were born in the early 1960s but it was really the introduction in 1964 of the 360 Series by IBM and the concept of Multi-Access Computing (MAC) that real-time processing became possible. By the end of that decade, organisations, such as banks, began to use terminals.

Since 1970, the whole shape of computing has changed, first with the mini- and subsequently the microcomputer. Now, companies such as STC, for example, have about 2000 terminals of which nearly 1000 are normally connected to one mainframe computer but with facilities to switch to other machines should the need arise. Information technology is now the buzzword and IBM has introduced the concept of document interchange architecture. Information is the driving force and the telecommunications network is the vehicle for the computer.

The communications manager who has traditionally been concerned with analogue audio circuits is now confronted with a whole new technology involving high-speed digital circuits. At the same time, the whole public switched network is being digitised. Digital speech circuits using Pulse Code Modulation (PCM) and electronic switching involving purely digital techniques are replacing analogue systems as fast as the manufacturers can produce them and administrations throughout the world can accept them. Meanwhile, IT, through data communication, is demanding higher rates of information transfer world-wide. They must go forward hand-in-hand to provide the integrated services of the future.

8: Installation of a LAN at St Thomas' Hospital

A V Stokes

St Thomas' Hospital
London
UK

As part of the implementation of their strategic plan, based on a distributed system of processors, interconnected by a communications network, an Ungermann-Bass Net/One (Ethernet) has been installed at St Thomas' Hospital. This paper describes the development of the strategic plan, the reasons for the selection of that particular equipment and the experience of testing and installing it in a production environment. Details are given of the tests that were carried out and the problems that were discovered in connecting various equipment to the network. The paper concludes that the network equipment is appropriate for the environment and it is currently in production use.

A V Stokes

Dr Stokes is Director of Computing at St Thomas' Hospital, London, and King's Fund Fellow. He obtained a BSc and a PhD from University College London. After working at GEC-Computers on language design, he rejoined the University at the Institute of Computer Science where his main work was concerned with computer networks (especially the link to ARPANET). He joined The Hatfield Polytechnic in 1977 as a Senior Research Fellow and took up his present post in 1981. Dr Stokes is a Fellow of the British Computer Society and a Member of the Association for Computing Machinery and the Institute of Electrical and Electronics Engineers' Computer Society. He has published nine books and over 40 papers on various aspects of computer science, particularly in the field of computer networks and videotex, and has lectured professionally, both in the UK and abroad.

Installation of a LAN at St Thomas' Hospital

Introduction

As part of the strategic plan for computing in West Lambeth Health Authority, a Local Area Network (LAN) has been installed in St Thomas' Hospital. This paper describes the strategic plan, the selection of the equipment required, the installation of the LAN and the future plans.

Background

Following the Flowers Report in 1966, a committee was set up at St Thomas' Hospital to look at the possible use of computers in the hospital and this committee recommended the setting up of a new department within the hospital for this purpose. The Department of Health and Social Security (DHSS) supported this idea and, in 1969, the ADP Project Team was formed to conduct a feasibility study.

In 1971, St Thomas' became part of the 'Coordinated Hospitals Project' in which four hospitals (Addenbrookes' Hospital Cambridge, Charing Cross Hospital, University College Hospital and St Thomas' Hospital) were provided with the same computer with the intention of being able to exchange large amounts of software. The computer chosen was the Xerox Data Systems' Sigma-6 and it was installed at St Thomas' in late 1973.

Over the next few years, a large number of applications were written for the computer. Two major areas were covered — patient administration and pathology laboratories — as well as a number of smaller ones, such as nursing sickness and absence recording.

In the late 1970s, the Sigma-6 was coming to the end of its useful life and a study was commissioned from external consultants to recommend the future of computing at St Thomas'. They produced a strategic plan for computing which recommended a distributed approach to computing.

The strategic plan

A hospital may be looked upon as a number of semi-autonomous units which have considerable internal needs for computing and communication but with relatively few links to the outside. For example, from the point of view of a user, a pathology laboratory will receive a specimen, then produce a report. In fact, this report is the result of extensive work, much of it involving computers (for example, a large proportion of laboratory equipment is connected to a computer). Thus a hostpital may be modelled as shown in Figure 1.

There are two obvious ways of implementing such a model on a computer. First (classically), each separate activity may be a separate process on a large computer and these may be interconnected by

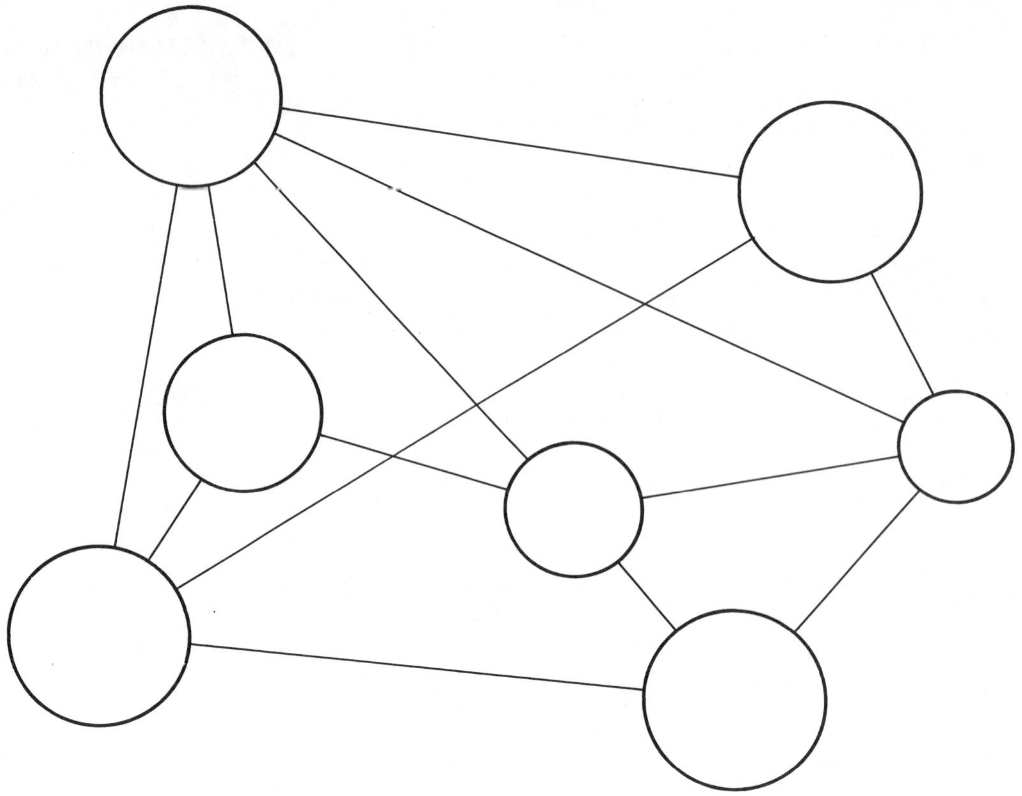

Figure 1: Model of a hospital

interprocess calls and shared files. Secondly, each separate activity may take place on a different processor and these processors are linked by a communications network. It is the second of these approaches that was adopted by the strategic plan. There were various justifications for this approach. In particular, it allowed an incremental changeover, followed by incremental growth. Once the transition to a distributed network had occurred, there would be no need to replace a mainframe once every seven to 10 years — the various computers on the network could be replaced gradually, thus spreading out the capital costs.

Furthermore, with such an approach, it is probably easier to integrate other machines into the system. Frequently, users within the hospital are able to purchase computers with, for example, research funds and then wish them to be connected to other systems (in particular to the computer containing the patient master index). There is often little, if any, control over the computers chosen and so a highly flexible approach is needed; this can be provided by adopting the distributed strategy.

It was realised that the proposed strategy had some disadvantages. In particular, the technology was considerably newer and problems could thus be envisaged. Secondly, it was likely that the overall costs would be higher (although Grosch's Law may not hold exactly, the basic philosophy is still true). Nevertheless, the advantages seemed to outweigh the disadvantages.

The strategic plan proposed that, at least in theory, each application should reside on a separate processor and that these processors should be interconnected. In practice, of course, it was realised that a number of different (but usually related) applications were likely to share the same machine but this did not detract from the general principle. However, a number of problems were evident, arising from the (semi-)autonomy of the 'owners' of these applications. First, the data needed within the hospital was distributed over a number of processors not centrally controlled. Thus it would be possible for data to be unavailable at critical times. Also, the method of accessing data would vary from machine to machine (or, from the user's point of view, from application to application).

In order to overcome these problems, the strategic plan proposed that there should be a particular system on the network providing any services which needed to be shared among the users of the network; this was called the shared services system. A number of functions were identified for the shared services system. First (and, probably, most important), it was necessary to provide a general enquiry service for all the commonly used data. This was designated the 'bulletin board' and was intended to provide a uniform method of access to all the data required for the day-to-day running of the hospital. Clearly, different users (or classes of user) had different requirements for access to data and the bulletin board therefore had to implement a high degree of security. It was decided that the information providers would not, at the time of providing the information to the bulletin board, specify to whom it was addressed; this would be done as a management function at an earlier stage. Thus, for example, it may be decided that all information from the nursing system sent to the bulletin board can only be accessed by nurses but no mechanism is provided for restricting access to information to a particular nurse.

The second shared service does provide that facility. It is known as the 'message board' and is a type of electronic mail in which messages must be addressed to specific recipients. The 'message' can, in fact, be the same data as was sent to the bulletin board. If a message is to be sent to a number of recipients, one copy must be sent to each recipient.

The third service is called the 'logon server' and is intended to provide a mechanism for controlling access to the network and for making the appropriate connections. With a user of a single computer system (which may, in fact, comprise a number of different processors), it is usual to 'log in' once only, then to access the particular facilities required. The user is unaware, if it is a multiprocessor system, on which processor his process is running, nor does it make any difference. Simply because a network is more loosely connected and the processors are likely to be dissimilar, this seems no reason to complicate the user interface and so it was decided to implement the logon server to emulate an interface that would be seen on a single machine. That is, the user logs into the network and not to individual machines within that network. To some extent, it is performing the functions of a Network Access Machine *(STO1)* but at a considerably less sophisticated level. More to the point, it is able to provide access control and security. Various other functions have been identified for the shared services system but these are not relevant to the present paper.

A final component of the strategic plan is the communications medium. Not only was it necessary to implement a LAN within St Thomas' Hospital but it was also necessary to provide links to other hospitals within the district, thus requiring interfaces to wide area networks.

A schematic of the strategic plan is shown in Figure 2.

Selection of equipment

Following the drawing up of the strategic plan and its acceptance by the hospital, work proceeded on implementation. The first stage was the selection and procurement of equipment. Two major items were required — the shared services system and the network itself. A second consultants' report was commissioned and a steering group set up, consisting of the consultants together with representatives from St Thomas'. This group made various recommendations which were approved by the District Management Team and work began on the procurement of equipment.

It had been decided that the communications network would be based on the Ethernet *(STO2)* for a number of reasons. Ethernet was a relatively well-proven technology and had reached the stage of a marketable product that could be bought 'off the shelf'. The transmission medium is passive and is also well shielded, both necessary attributes in a hospital environment. It conformed to a formal standard *(STO3)* which had been proposed as a national standard (within IEEE's Project 802). Finally, there was the possibility of an upgrade path from baseband to broadband transmission should this be required in the future.

The DHSS was approached and asked to fund a technical study into the implementation of an Ethernet and this was approved in January 1982. An order was placed for a Net/One system from Ungermann-Bass and this was delivered in March 1982.

The Net/One system

Net/One presents the user with considerably more facilities than the basic Ethernet. At the lowest level, the system consists of a cable to the full Ethernet-2 specification *(STO3)* allowing data transfer at 10 Mbit/sec. A transceiver is required to transmit data and this is clamped directly onto the cable (since

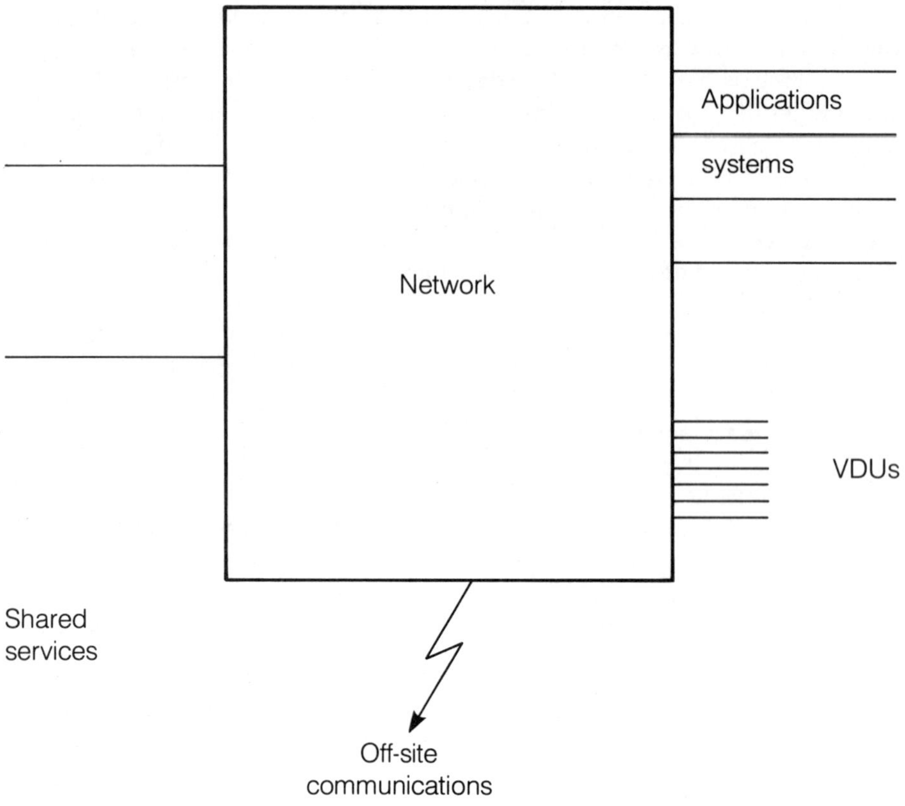

Figure 2: The strategic plan

the transmission medium is passive, transceivers may be added or removed while the network is in use). This is connected, in turn, to a Network Interface Unit (NIU).

An NIU consists of a number of boards, each of which contains a Z-80 processor and 64 Kbytes of Random Access Memory (RAM). This is interfaced to an I/O board which may handle, for example, four serial (RS-232) and two parallel connections. In fact, a variety of different I/O interfaces are available. In the NIUs, which were bought for the first phase of the evaluation (NIU-2As), there was space for four boards to enable each NIU to handle a maximum of 24 devices. A later version of the NIU based on the Fujitsu Ethernet chip set (NIU-1/50) became available in 1983 and was evaluated during Phase II. The NIUs implement the basic Ethernet protocols but, in addition, provide the user with higher-level facilities enabling, for example, connection to a named device elsewhere on the network. The NIUs also guarantee an error-free transport service.

Various types of service may be obtained through an NIU. The most common is access to the command interpreter which allows functions such as CONNECT to another port, DISCONNECT and EXAMINE the status of a connection. A user who has access to the command interpreter has access to all the facilities through the network but has to issue specific commands.

A second service is that of the Permanent Virtual Circuit. The network can be configured so that one port appears to be permanently connected to another. The user has no need to (and, indeed, cannot) issue network commands.

A further facility is a demanded circuit and this is discussed later.

Phase I — initial systems testing

The technical study for the evaluation of the network was intended to take 18 months to complete but, for various reasons, was extended by a further six months. Thus it terminated at the end of March 1984. It consisted of two phases. The first was intended to be a thorough test of the system (both hardware and software) in an experimental environment. When the system had been proved to be acceptable, it would be installed and then used in an operational environment.

The equipment purchased for Phase I consisted of one Network Configuration Facility (NCF-1) using twin 8-inch floppy disks as their storage medium, five NIUs providing a total of 102 ports (most were RS-232 although there were a few parallel ports) together with 500m of Ethernet cable with all the appropriate fittings. In fact, the equipment delivered did not quite conform to the order. In particular, the NCF-1 had been dropped during transit and an NCF-2 (with a hard disk instead of one of the floppy drives) was supplied until a new NCF-1 could be shipped from the US. The equipment was powered up and a fault discovered on one NIU. This was due to some chips on one board becoming loose during transit and so the fault was quickly rectified.

The initial tests were extremely simple and consisted of connecting a terminal to each (serial) port on each NIU to ensure that the command interpreter could be accessed. The next set of tests involved two VDUs connected to different ports. Since testing was intended to be exhaustive, as many combinations of ports as possible were checked (although it was decided that testing the 5000 or so possible combinations was too extreme) and as many facilities as possible were used. Thus, for each pair of ports, tests were conducted with both ports at the same baud rate (over all possible baud rates), at different baud rates, at the same and different parities, etc.

In general, these tests were successful but a few problems did materialise (thus demonstrating the need for such exhaustive tests). In particular, a fault was discovered on the SIXPAC boards (ie those supporting six serial ports rather than the '4+2' — four serial and two parallel — boards). This manifested itself by the echoing of a spurious character when local echo had been selected and this was found to occur at regular intervals, equal to the input buffer size. This was clearly due to a software error concerned with buffer pointers.

Having made successful connections, tests were performed on disconnecting and these were successful, although a problem was discovered later. For each port, a sequence of characters is defined to be that which performs the disconnect function and this must clearly be chosen to be a sequence that would not normally occur. For most devices, the sequence '@end' was chosen. While this was found to work when typed in via a terminal, it was later discovered that it did not work when sent by a machine, unless the machine paused between characters! Again, the explanation seemed obvious.

The third series of tests were concerned with flow control. Net/One allows ports to be configured with one of a variety of flow control mechanisms such as XON/XOFF or RS-232 line signalling. While the network appeared to perform correctly with the various types of flow control, a problem was discovered when connecting the network to the mainframe computer, the Sigma-6. This arose since the Sigma did not implement XON/XOFF flow control but, in addition, although the documentation suggested that it did implement line signalling, this was found not to be the case. The problem was eventually solved by Honeywell (which maintained the Sigma) supplying a 'patch' to the system software to implement RTS/CTS line signalling correctly. A further problem was discovered in this area; when a port was configured without flow control and data was sent too rapidly to the NIU, rather than discard characters, it appeared to write beyond the end of the buffer and 'hung'. This problem was solved in a later software release.

Having established that the system appeared to perform satisfactorily (with a few problems) under normal circumstances, tests were conducted to simulate failure conditions. For example, while in the middle of transmitting data, an NIU was powered off. In all cases, the appropriate recovery procedures were effected. In the test just mentioned, the NIU, on being powered on, performed the normal power-up procedure (self-diagnosis, followed by requesting the NCF to down-load its configuration) with all circuits prior to the 'crash' being terminated. Clearly in an operational environment, this is not convenient and the users would have to reconnect to the particular destination; however, the alternative of some third part (or the network itself) performing the reconnections would imply more centralised control than presented by Net/One. In any case, it is expected that such occurrences would be rare.

During the initial tests, various problems were discovered, as outlined above. Some hardware faults were noted and one took a long time to clear up since it seemed to be due to a combination of faults. The symptoms were that some boards of an NIU kept ceasing to operate. A number of different solutions were proposed — for example, the Ethernet cable is marked at 2.5m intervals and NIUs should be connected to a mark but, in our case, this had not been done. On doing it, no difference was noted (except temporarily). Another proposal was that it was caused by the Ethernet cable being looped around on a drum; it was unwound without effect. Eventually, the problem was solved by replacing the transmitter board and a new update to the network software.

Various faults were noticed in the software and, during Phase I, about eight were serious enough to be reported to the supplier. All were remedied in various later releases of the software but, in some cases, a temporary patch was supplied prior to the next release.

To summarise, the number of problems found during Phase I were much as could be expected. More faults were found in the newer devices (particularly the SIXPAC boards) and a major problem was the dearth of debugging tools or other diagnostic aids. Nevertheless, by the end of 1982, Phase I was deemed to have been concluded and it was decided to implement Net/One as an operational network within the hospital.

Phase II — the operational network

All tests during Phase I had taken place in an experimental environment, namely in a corner of the computer room. Once it was clear that the product could operate successfully, the network cable was laid throughout the hospital (although a short piece of cable was retained in the computer room for testing purposes — for example, when a new release of the network software was received).

Although this sounded a straightforward task, a number of problems were encountered. First, the cost of laying the cable was quite high and considerably more than laying the normal multiway data cables. This was undertaken by specialist outside contractors. Secondly, despite measuring the route of the cable as accurately as possible, the cable did not reach its destination; this was attributed to the fact that, due to the rigidity of the cable, it was not possible to bend it and so significant amounts were 'lost' on corners. In fact, it was necessary to extend the cable by about 100m to compensate for this (bringing its total length to 600m which is more than the specification allows, although no problems have been noticed as a result).

In addition, choosing sites for the NIUs was not as easy as expected. Various criteria had to be applied such as the site having power available (13 amp mains), being accessible to computer centre staff while being inaccessible to the general public and being reasonably well ventilated.

Nevertheless, within a relatively short time, a number of NIUs were distributed around the hospital. Some were retained in the computer room since all extant wiring terminated there and it seemed pointless to rerun data cables (except where this would be cost effective — for example, where the use of line drivers could be obviated).

This implementation was phased to fit in with a programme of installing terminals on wards and the provision of a 24-hour, seven-day enquiry service on the Sigma-6 and so, as the network was extended, the major use was from these terminals accessing the enquiry service.

It was also used for various other purposes such as providing a dedicated link (a permanent virtual circuit) between a terminal on one ward and a microcomputer in the Department of Medicine. Thus, as far as the operation of the hospital was concerned, implementation of Phase II provided, almost immediately, additional services. As far as the technical study was concerned, considerably more work was undertaken. In addition (and in parallel) work continued on implementing the strategic plan.

Perhaps the first point to be noted is that, as the network was extended and more terminals added, these terminals were from a number of different manufacturers and had significantly different characteristics. The terminals included two models of Cifers, the TVI-925, Kokusai KDS-7362 (which is very similar to the TVI-950) and hardcopy terminals such as a DECwriter III and a Diablo. In all cases, no problems were encountered.

Similarly, a number of microcomputers have been attached (through RS-232 connections) and, once again, no problems have been encountered. This is to be expected, since the network has been used, in all these cases, merely as a replacement for a hard-wired connection. Nevertheless, as mentioned above, one connection has been used in a live clinical environment on a 24 hours a day, seven days a week basis since the middle of 1983.

Connections to minicomputers have also proved trouble free. A Prime-300 computer was connected for early tests (but the machine was later sold). In addition, a twin GEC-4190 configuration (bought as the shared services system and described below) has been connected since autumn 1983.

Finally, two mainframes have been connected. The first of these is the Sigma-6 and the work on this has been discussed above. It is now fully operational and few problems have been encountered. The ward terminals access it using a facility that was introduced into Net/One in 1983, known as a 'demanded circuit'. This is similar to a permanent virtual circuit but is not a permanent connection. A port is configured for a demanded circuit to a particular destination (in this case, the Sigma) and, when any key is hit on the terminal connected to that port, a connection is made to the selected destination. The 'destination' is a network name and can therefore be a single port or a group of ports, in which case, the first free one is used. Thus the network may be used as a terminal multiplexor and, at the time of writing, about 50 terminals have access to eight Sigma ports. In order to implement this satisfactorily, it is necessary to impose a relatively short time-out on these ports since, otherwise, they would rapidly become blocked.

The second mainframe temporarily attached to the network was an ICL 2955. A large number of problems were encountered, mainly due to the fact that this machine used synchronous terminal protocols. Experiments to use the Net/One to connect the mainframe and its cluster controllers (DRS-20/33s) were partially successful. Connections were established but down-loading the DRS-20s did not take place. The experiments were terminated but may be restarted at a later date.

Similarly, access to the ICL machine using asynchronous links proved difficult, despite the installation of an ICL product known as a CMS-1 board. The major problems were concerned with flow control and it only proved possible to obtain reliable communications using 300 baud links. Again, these experiments were terminated but may be restarted.

During Phase II, the network software improved and, in particular, a debugging tool known as the Data Link Monitor (DLM) became available. A sample of its output is given in Figure 3. This enables

Data Link Monitor 1.0 *Freeze* Monday February 20, 1984 17:08:22

NIU	Send pkt	Recv pkt	Send pkt/s	Recv pkt/s	Colli-sions	CRC errors	Active boards
1156	1401344	1426673	0.5	0.5	283	0	AB
806	13395792	13434885	1.5	1.5	1223	0	ABCD
1947	3860399	3860624	10.4	10.3	369	0	ABCD

Board	Send pkt	Recv pkt	Send pkt/s	Recv pkt/s	CPU util %	Response time ms	Net/One service
806A	455463	482657	0.2	0.2	6.8	16	VC
806B	1895347	1923552	0.1	0.1	6.5	16	VC
806C	1726702	1761164	0.1	0.1	6.5	16	VC
806D	9317526	9342861	0.9	0.9	18.8	16	VC

Command Menu (Type ? for help)
1) Add line 4) Freeze display
2) Delete line 5) Unfreeze display
3) Clear data 6) Exit

Select command number: 1
Address of niu or board: 806D
Select command number: 4
Select command number:

Figure 3: Typical DLM output

traffic to be monitored and has proved extremely valuable in pointing out a fault which might otherwise not have been discovered for some considerable time.

On first using DLM, it was noted that the number of packet collisions was very high and this was traced to incorrect termination of the cable. Once rectified, the number of collisions dropped to almost zero (since, at present, the amount of network traffic is only a very small proportion of the total capacity).

A number of hardware and software problems were encountered in Phase II. The number of hardware faults (mainly board failures) was considered reasonable and the revenue costs of the system were relatively low (first-line maintenance is carried out at the hospital). Most of the software errors were still outstanding at the end of Phase II.

The shared services system

As mentioned above, the strategic plan also called for a computer system to provide shared services on the network. An operational requirement was produced and over 50 companies requested copies. Eventually, four companies were shortlisted and, after careful evaluation, it was decided to purchase a twin GEC-4190 configuration with two 275 Mbyte disks on each machine, an intercomputer link and a watchdog timer. The computers are therefore now interconnected and each is connected to a number of NIUs, thus allowing relatively fail-safe operation. At present, some systems are still being developed on the machines and they are not in operational use.

A number of systems are being developed and two of these are of particular interest. They are the bulletin board and the logon server (both described above). The former is being implemented using a relational database system, RAPPORT, and a prototype is running for patient administration data. A prototype logon server has also been implemented and this makes use of the demanded circuit facility of Net/One. These two systems are expected to be implemented in the near future and work is proceeding on the other shared services.

Summary and conclusions

This paper has outlined the implementation of a strategy at St Thomas' Hospital for implementing a distributed, heterogeneous network and, although implementation is still proceeding, the results so far have been highly encouraging. The decision to use an ETHERNET-type LAN has been more than justified, although the teething troubles that were perhaps to be expected with a relatively new product have proved troublesome. However, the suppliers have given considerable support and there are now no reservations about its use in an operational environment.

Having implemented the network and associated systems, it will prove very much easier to interconnect the wide variety of computer systems that are available (and will become available) both in the hospital and in the district.

Acknowledgements

The author would like to acknowledge the contribution of the DHSS who funded the technical study on which part of this article has been based and would also like to thank the various people who have contributed to the work on the network in St Thomas' Hospital, including the staff of the Sharpey-Schafer Centre and the members of the Network Steering Group for their valuable suggestions.

Appendix figures

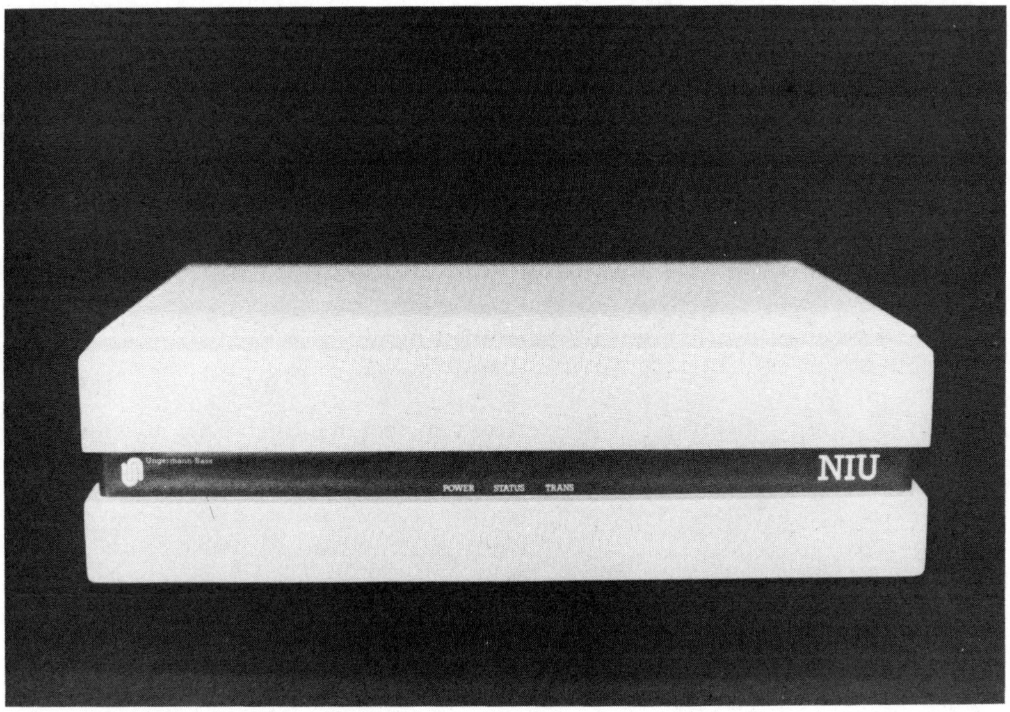

Appendix Figure 1: The Ungermann-Bass NIU-1/50

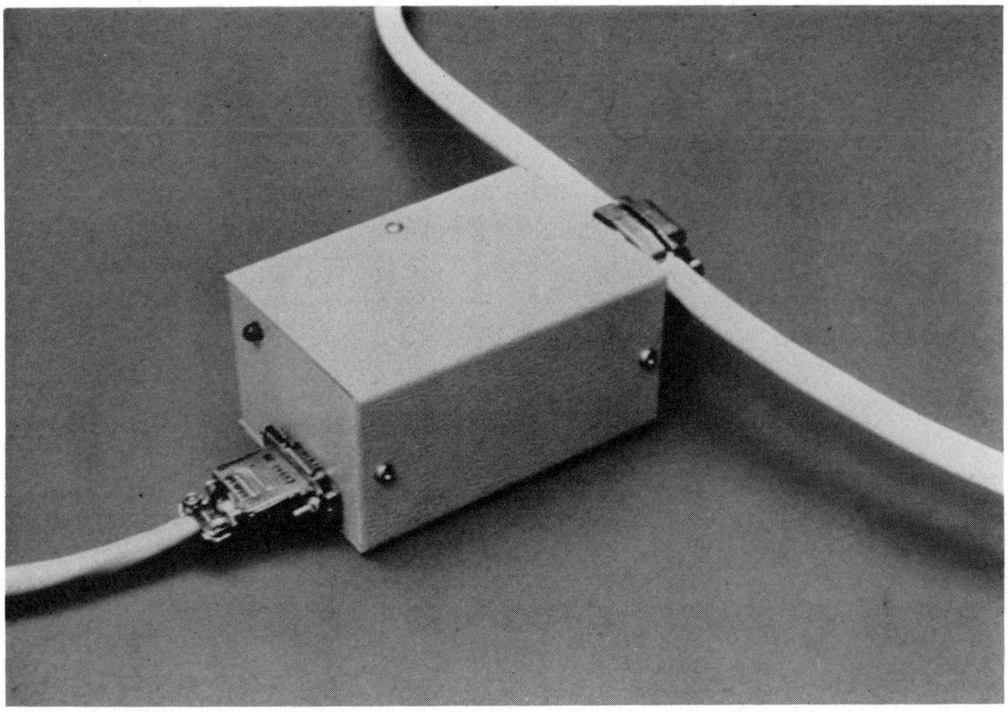

Appendix Figure 2: A Net/One transceiver connected to the ETHERNET

9: Open Systems Interconnection — present state and strategies for adoption

J Tucker

Logica UK Ltd
London
UK

Open Systems Interconnection (OSI) is an ambitious standardisation programme which is being undertaken by ISO and CCITT, in collaboration, with the objective of producing standards for protocols that will make a wide variety of distributed applications possible. Within the next 18 months, technical stability should have been achieved in almost the whole of OSI. Then the development of OSI will be seen as a very successful and immensely valuable exercise.

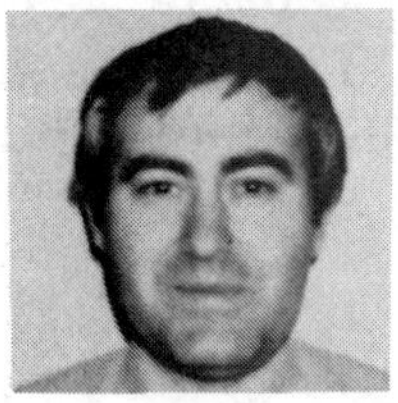

J Tucker

Mr Tucker is a Principal Consultant at Logica UK Ltd where he specialises in communications, computer networks and related standards. He has worked in this field for more than 10 years and his experience includes early studies of ARPANET, a design study of the British Steel Network COTEN, implementation of the European Informatics Network switching node, and functional specifications for a variety of computer networks thereafter. Since the mid-1970s he has been active in international standardisation, first in HDLC and subsequently in OSI, and he is currently convener of the ISO working group (TC97/SC6/WG2) responsible for the Network Layer. He read Mathematics at Christ's College, Cambridge, where he also gained the Diploma in Mathematical Statistics.

Open Systems Interconnection — present state and strategies for adoption

Introduction

Open Systems Interconnection (OSI) is an ambitious standardisation programme which is being undertaken by ISO and CCITT in collaboration, with the objective of producing standards for protocols that will enable a wide variety of systems to communicate in support of a wide variety of distributed applications. Six years of intense activity are at last beginning to bear fruit. Already some OSI protocols are technically stable (and many more will become so during the next year or 18 months) although official publication as international standards and CCITT recommendations will take at least another year. By mid-1985 therefore, stable protocols will have been specified to meet the objectives of OSI for a wide range of environments. However, this reduction of protocol variety will not yet fully guarantee that a system complying with OSI protocol standards will be able to communicate with any other complying system; a small number of variants will exist and market forces will be needed to select between them. Nevertheless, the programme must be judged a considerable success.

The information technology industry will not wait until mid-1985 before it begins to exploit OSI. Already a number of coordinated efforts are being made to anticipate the outcome of standardisation in order to gain early benefit from the openness of OSI. The ideal of a world-wide 'anticipation' plan for OSI is not attainable, but a national plan is under development — namely the Intercept Strategy of the Department of Trade and Industry. Attempts are being made to extend this plan into a harmonised standards policy for the European Community.

Finally, some words of caution. OSI is a programme of development of specific protocols which are designed to have very wide applicability in data communications: it is not, however, the solution to all known problems. OSI is specific; it is not something that anyone can simply pick up and reinterpret or modify; nor can sections be extracted at will. If this does happen the result will not be OSI and should not be described as such. Furthermore, the protocols, and even the Reference Model, may not be appropriate in wider contexts. In particular, their applicability to voice and wideband communications deserves careful examination: it should not be a foregone conclusion.

Ultimately it is hoped that OSI will form part of a wider set of standards for communicating systems which not only embrace these and other forms of communication, but which also extend upwards to include a number of distributed applications. Telematics and message handling services may be seen as early examples of such systems.

The nature of OSI

What OSI is

It should be evident to any reader of the computer press, journals, conference proceedings and manufacturers' publicity material that OSI is widely accepted as the basis for communications standards of the future. To a large extent this testifies to the solid progress made in international standardisation, but there is a danger, in the haze of publicity, of losing sight of precisely what OSI is and what has indeed been achieved.

OSI is an international programme of standardisation, conducted by ISO and CCITT in collaboration, the result of which will be a set of standards for communications protocols, divided into layers. A system implementing a subset of these protocols and which complies with conformance statements in the standards for the various layers will be termed an *open system*. The objective of the standardisation programme is that, within very broad groupings, open systems will be compatible in the sense that they can communicate with one another to provide communications support to a wide variety of distributed applications.

It will be seen that, in order to give structure to such a mammoth standardisation programme, it is necessary to:

- Partition the entire set of communications functions
- Allocate them to layers
- Provide functional specifications for the protocols for each layer.

The agreements on these structural matters are documented in standards called the OSI Basic Reference Model and Layer Service Definitions. They are made standards in order to freeze the specifications, not for external consumption. They are of importance only to those developing OSI protocol standards: there is no sense in which systems can claim to comply with them.

Who is standardising OSI?

OSI is under collaborative development by ISO and CCITT. In ISO the standardisation of OSI falls within the scope of Technical Committee (TC) 97 and, until now, has been developed mainly in subcommittee (SC) 16 with the lower three layers being the responsibility of SC6. At the time of writing, however, TC97 is being reorganised and, while the exact effect of this is obviously not known, the current proposals are that OSI should be reallocated to SC6 and a new SC21. The intention is to extend the responsibility of SC6 to include the Transport and possibly also the Session Layer and for SC21 to gain some work items from SC5 as well as SC16 (which is then expected to disappear).

In CCITT the OSI standardisation activity has taken place within Study Group VII, and the level of agreements reached there in 1984 will be documented in various X.Series Recommendations. CCITT's method of working is quite different from that of the ISO in that it is organised into four-year study periods at the end of which any level of agreement that may have been reached is published in the Coloured Book of the period. ISO, in contrast, does not publish an international standard until complete technical agreement is reached on its contents: moreover, notes that certain items require further study are not permitted.

Inevitably the pace of international standardisation is slow since so many countries and organisations, all with different interests, are involved. OSI has already been in the development phase for over six years and only now are firm agreements being established for some of the layers. It has thus been a long process which is only just beginning to produce results.

In this context it is perhaps useful to clarify the part played by the European Computer Manufacturers' Association (ECMA) in the development of OSI. ECMA is a major contributor to ISO and may justly claim credit for much of the progress that has been made. ECMA's method of contribution to OSI is, by and large, to agree ECMA standards or technical reports in-house and before submitting them to ISO. After careful consideration, influenced by comments from ISO member bodies and CCITT, ISO modifies the standards. Finally, when ISO reaches technical agreement, ECMA, as a matter of policy, brings its own standards into line with ISO. An ECMA standard or technical report is thus an important and influential contribution to ISO. It is not an authoritative statement of OSI, but the view of one significant interest group that has an efficient organisation for reaching and publishing agreements.

As an agreement between manufacturers, an ECMA standard on OSI may also have a role in a manufacturer's migration strategy towards OSI, although this will depend on the progress being made within OSI by the subject in question. Migration strategies are further discussed under the heading 'The adoption of OSI'.

What OSI is not

Before discussing the present state and likely future of OSI, it is important that some common misconceptions and mis-uses of OSI ideas and terminology should be dealt with. The computer press and journals abound with vague and often misleading references to OSI, some of which demonstrate misunderstanding, while others suggest a desire to be fashionable rather than any real commitment to general intercommunication.

Firstly, OSI is not a general-purpose, descriptive tool that will provide an understanding of other sets of protocols. Occasionally the OSI Reference Model is (mis-)used in this way, but this only leads to confusion. The Reference Model is neither general purpose nor descriptive: on the contrary it is a special-purpose and prescriptive tool which makes specific allocations of functions to layers, thus deliberately constraining (ie part-specifying) OSI protocols. Other network architectures may partition functions differently, in which case the Reference Model fails to describe them.

Secondly, the OSI Reference Model is not a licence to apply the term 'open system' to another system, some of whose protocols may bear some relationship to some parts of the OSI Reference Model. It is distressing to see publicity of the form:

> *The X system/protocol/architecture is based on the OSI Reference Model. It uses layers A, B and C of OSI but the other layers are not needed. The protocol for layer A contains ....* [is not the ISO standard for the layer and probably does not perform the functions the ISO protocol performs].... *It offers the ability to communicate with ....* [other identical systems, but not OSI open systems at all].

As a general rule, any claims about open systems or OSI that do not refer to specified ISO (or CCITT) *protocol* standards should be viewed with some caution.

The structure of OSI

The objective of OSI is the publication of standards for communications protocols in order that systems obeying them can interact with each other usefully, to support a wide variety of distributed processing needs over a wide variety of communications media. The complete set of functions required to enable this interaction is vast and complexly interrelated. Some groups of functions are specific to particular application needs, such as file transfer; some are specific to particular communications media while others are always needed. Some groups of functions are independent alternatives while some rely on others (for example file transfer relies on support for end-to-end communication across multiple networks).

If any progress at all is to be made in the face of such complexity and if a set of protocols is to be built which is flexible enough to evolve in order to satisfy changing user needs over new technologies, then the complete set of functions must be split into groups of functions which can efficiently and sensibly be performed together, and the job to be done by those groups must be precisely defined. There are therefore three necessary stages of development:

1 The organisation of functions into groups.
2 The definition of precisely what those groups are required to do.
3 The detailed specification of the functions themselves in protocol operation.

In OSI these three stages are clearly distinguished, although they have not been performed entirely sequentially. The three stages are:

1 Architecture.
2 Service definitions.
3 Protocol specifications.

Architecture

This first stage, which is almost complete, is the organisation of functions into groups, ie *architecture*. The functions are organised into seven well known layers, as defined in the OSI basic Reference

Model IS 7498. The layer structure is a refinement of the basic idea that the way in which a particular distributed processing function is implemented (file transfer for example) should not depend on the fact that the two systems performing it might, by chance, be linked by an X.25 and an X.21 network. Furthermore, methods of interworking across an X.25 and X.21 network should not be defined with file transfer specifically in mind. If on the other hand the two tasks are approached independently, both can have general applicability, even to environments not as yet foreseen. When this approach to OSI was adopted, it was generally agreed, though not without considerable discussion, that seven layers would be most appropriate. These layers are illustrated in Figure 1 and their tasks outlined in Figure 2. An important point to note is that the layers sit above one another in the sense that each particular layer relies on the operation of those below. They are not alternatives: each is indispensible.

As previously noted, the architecture is well defined in IS 7498. There are two other dimensions of classification, however, that are not yet completely defined, both of which are refinements of the layer structure. The first refinement reflects the fact that the tasks of some layers may depend upon the nature of the application or upon the environment. File transfer mechanisms, for example, will be needed sometimes but not always, as will virtual terminal operation and remote job transfer. Similarly, the details of the operation of the Network and lower layers depend on the communications media; where networks are interconnected those layers will use different mechanisms — to meet the same needs — in different places. This type of refinement of the layer structure is most significant in the Network and Application Layers.

The other refinement of the layer structure is for management. There is clearly a need for some functions that are not specific to particular communications but which are there to control and monitor the use of system resources, for example:

- Updating routing tables or addressing directories
- Collecting statistics on traffic and performance
- Taking equipment out of use and reintroducing it.

OSI management is still at an early stage of development and not unnaturally the highest priority has been accorded to developing the standards for 'normal' operation.

Service definitions

Once the broad groupings of functions have been determined, the next task is to define precisely what it is they have to do. This is carried out by *service definitions*. The key principle in this area of OSI development is that of the independence of layers, or perhaps more accurately the independence of layer protocols. The idea is that no protocol should rely upon knowledge of the details of any other protocol, but should instead be designed to meet functional specifications for its layer independently of how the protocols for other layers are designed to meet their own functional specifications. These functional specifications are being produced in the form of standard service definitions. The service definition for layer (N) is a definition of what all layers up to and including (N) provide, as a 'service', to the layer above. The functional specification for the protocol for layer (N) is therefore the (N)-service definition, which states what the layers up to (N) must provide, together with the (N-1)-service definition, which states what services the layers up to N-1 make available to the (N)-protocol.

Service definitions are necessarily written in abstract terms, since they must not unduly restrict the means of implementing internal communications used within an open system. It is not even a requirement that an interface corresponding to a layer boundary can be identified within an open system, so long as the protocols seen externally are OSI protocols. Service definitions, then, talk of establishing connections, of passing 'primitives' with 'parameters'; they do not talk in terms of the values of bits in the second octet in the header — these are matters of protocol. However, this does not make them any the less precise and in fact the use of an abstract style is essential if the requirements themselves are to be distinguished from the manner in which they are to be met.

Service definitions are standards for standardisers, namely those developing the standards for protocols. Real systems may comply with protocol standards; it is meaningless to talk of their complying with service definitions.

Protocol specifications

The third and final stage is the development of the *protocol standards* themselves, to meet the service

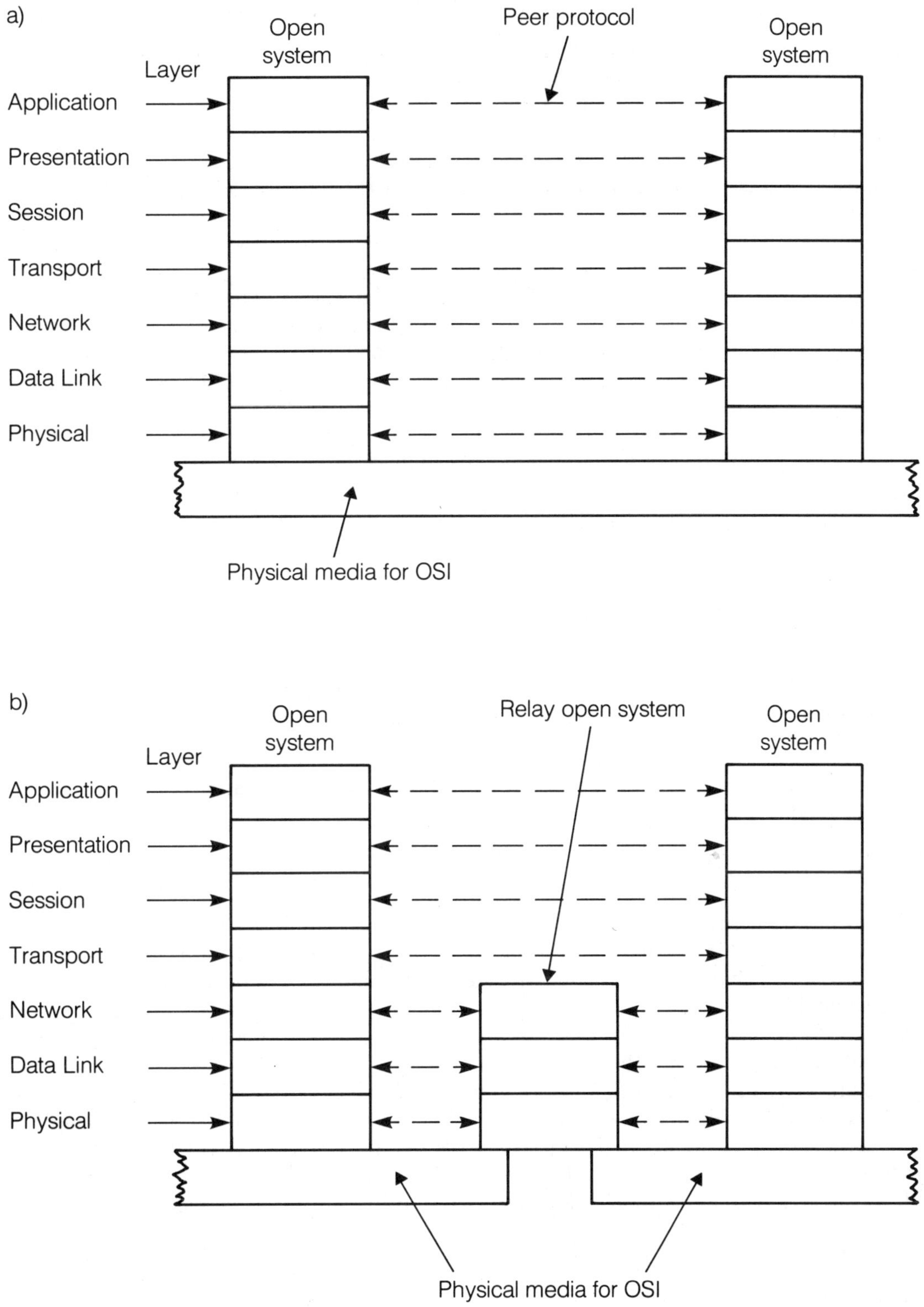

Figure 1a: Seven-layer reference model and peer protocols
Figure 1b: Communication involving relay open systems
Figure 1: The seven-layer reference model taken from IS 7498

Application Layer:	The window to the OSI environment. Support for particular common requirements, eg: • File transfer, access and management • Virtual terminal • Remote job transfer and manipulation.
Presentation Layer:	Identification/negotiation of the syntax used for data transmission.
Session Layer:	Organises and controls the dialogue: two way simultaneous/two-way alternate operation, establishment of synchronisation points, resynchronisation and graceful termination.
Transport Layer:	Enhancement where necessary of quality or cost-efficiency of network service by end-to-end error recovery, multiplexing etc.
Network Layer:	Provision of a standard end-to-end service across communications media, singly or in combination, that is independent of the characteristics of those media (except for quality of service).
Data Link Layer:	Provision of reliable, blocked transfer of data across the physical medium between adjacent systems, in many cases with flow control.
Physical Layer:	Provision of mechanical, electrical, functional and procedural means to implement bit transmission across the physical medium.

Figure 2: The functions of the OSI Layers

definitions. It is at this stage that the details of formats, encodings, timings, allowable sequences etc must be specified so as to ensure (as far as possible) that two implementations of the same standard will interact as desired.

OSI and real networks

In a desire for completeness, the OSI Reference Model specifies an architecture that starts right at the bottom with the physical medium connecting two systems and then builds the layered structure on top of it, right up to the Application Layer. However, the OSI standardisation bodies will not be developing all the future standards for communications interfaces and networks, and so it has to be recognised that real networks will not necessarily be developed to use OSI protocols and may not even correspond in their allocation of functions to the Reference Model. Today, the most commonly-referenced network procedures are those of X.21 and X.25 in the public domain, with ISDN on the horizon and the IEEE 802 standards for local area networks. None of these corresponds exactly to OSI, although X.25 comes close. It is not even desirable that OSI protocols should be applied right at the bottom of the architecture since that would put constraints on innovation in communications technology. What is needed instead, is a defined method by which a variety of low-level communications procedures can be accommodated in a world of interconnected networks across which OSI can be used and its benefits obtained. Such a method exists. For example, the supplier of a communications network determines the facilities it offers.

Those facilities can be represented in OSI by a service in the OSI sense of the term. Such a service is called a subnetwork service to distinguish it from the Layer 3 network service and to indicate that the real network in question is seen as constituting a subnetwork of the global collection of communications networks. Clearly, different subnetworks will in general provide different subnetwork services. In principle, for any subnetwork service (in practice for all the significant ones) one or more OSI network protocols will be standardised capable of 'bridging the gap' between the subnetwork service in question and the OSI network service. (The extent to which these protocols should be general as opposed to subnetwork-specific is an important Network Layer issue.) In effect, the idiosyncratic characteristics of particular networks will not be visible above the Network Layer, over which standard OSI protocols may operate. The Network Layer also has the task of interconnecting subnetworks in order to provide communications end to end: that again will not be visible to higher layers.

The state of OSI

Progress towards standards

The current state of most of the OSI standards is shown in Figure 3 which gives target dates for the following milestones in the progress of a standard:

1 Draft Proposal (DP): at this first level the document is assigned a number which will accompany it throughout its progress and is issued for ballot at subcommittee level.

2 Draft International Standard (DIS): when a DP has successfully completed its first stage of balloting and any necessary revisions have been made, it is issued for ballot as a DIS at technical committee level.

3 International Standard (IS): when the DIS ballot has been successfully completed, the document is issued by Central Secretariat as an IS.

The dates given in the figure are target dates by which time all the technical conditions for issuing the documents at the appropriate level should have been met: there are publication delays at DIS and IS level, significantly caused by requirements for translation.

The majority of CCITT activity on OSI is concentrated on the Reference Model (CCITT version Rec.X.200) and on:

- Network service
- Transport service and protocol
- Session service and protocol.

There is very close agreement between CCITT and ISO texts for these. CCITT is publishing its versions in the following structure:

- Rec.X.21*n* is the service definition for layer *n*
- Rec.X.22*n* is the protocol specification for layer *n*.

Achieving compatibility

The picture painted in Figure 3 is a very encouraging one. Within a year or 18 months from now technical stability should have been achieved in almost the whole of OSI (although the ISO Central Secretariat will take a little longer to publish the International Standards). The development of OSI will be seen as a very successful and immensely valuable exercise.

It is, however, appropriate to ask how far this development has gone towards achieving its original goal, namely the standardisation of protocols such that open systems using them can communicate to support distributed applications. The answer is that the original goal has very nearly been achieved. Protocols will indeed be standardised to support distributed applications: the only thing missing is that compliance to OSI standards will not by itself guarantee compatibility. A small number of protocol variants will be permitted by the standards, with the effect that inappropriate choices of variants could lead to incompatibility between systems. It is therefore necessary to identify the OSI protocol set selected if compatibility is to be ensured.

Protocol variants arise out of the conflict between generality and efficiency. An early view, or aim, of OSI was that there should be just one set of protocols (one per layer) that open systems would use, and that the uniqueness of this set would guarantee compatibility. However, in reality there is such a variety of requirements that such a unique protocol set would have been expensive to implement and inefficient to operate. The obvious solution was to invent alternative protocols and standardise a means of negotiating the variant to be used in any particular case. Different variants would be appropriate in different circumstances and gross inefficiency could thereby be avoided. Furthermore, not all the variants would need to be implemented in every system so long as one variant were made mandatory to serve as the default protocol in the event that preferences for other variants could not be satisfied.

STANDARD	1984	1985	1986	1987
Basic Reference Model	IS			
CL-addendum to RM	DP-------DIS	--------------IS		
Naming and addressing		DP----------	---DIS--------	---IS
Security in OSI		DP---	----------DIS-	--------------IS
Further applications layer standards*		DP-----------	---DIS--------	--IS
Basic class virtual terminal		DP---------	---DIS--------	---IS
Other virtual terminals*			DP-----------	-DIS----------
Remote job transfer and manipulation S&P	DP----	----------DIS-	-------------IS	
File transfer, access and management S&P	DP-------DIS	--------------IS		
Presentation service CO	DP----	----------DIS-	--------------IS	
Presentation protocol CO	DP----	----------DIS-	--------------IS	
Session service CO	DIS--------IS			
Session protocol CO	DIS--------IS			
Transport service CO	DIS--------IS			
CL	DP---------	---DIS--------	----IS	
Transport protocol CO	DIS--------IS			
CL	DP---------	---DIS--------	----IS	
Network service CO	DIS-----------	------------} IS		
CL	DP------DIS	------------}		
Network addressing	DP-------	---DIS--------	----IS	
Network protocols* CO	DP------DIS	--------------IS		
CL	DP------DIS	--------------IS		

Key:
* = Substantial area of work, which will lead to many standards; the dates indicated are for the earliest of these.
CL = Connectionless.
CO = Connection-oriented.
S&P = Service and Protocol.

Note:
The dates indicated are those predicted for the reaching of technical agreements and the completion of the necessary balloting; there will be a further, inevitable delay before the documents are published by ISO. Also the dates are targets and therefore may be optimistic.

Figure 3: Target dates for OSI standardisation (major standards)

This is precisely the procedure that has been adopted in the development of OSI, with considerable success. There are one or two areas, however, in which it has not proved possible to reach international agreement as to which variant should be the mandatory one. In those areas, therefore, compliance with OSI protocol standards will not by itself guarantee universal compatibility. The first example of this occurred with the five *transport protocols* (called 'classes') developed to provide the connection-oriented *transport service*. Any complying system must implement Class 0 *or* Class 2, but a system that implements only Class 0 cannot communicate with a system that implements only Class 2.

A similar difficulty, on a greater scale, has arisen with the standardisation of connectionless services and the protocols to support them. The basic reference model is built on the notion of connections at every layer. The operation of each layer is seen as having three phases:

1 Connection-establishment, where negotiations take place and resources are allocated.
2 Data-transfer.
3 Connection-release.

After a period of development the concern was expressed that, whilst connection-oriented operation was what was required in many circumstances, there were other circumstances in which simpler, more efficient protocols could be used. Transaction processing of an enquiry-response kind and datagram networks were cited in support of the argument, and there grew up a strong lobby wishing to extend OSI to encompass new 'connectionless' services, supported by simple protocols.

This is not the place to argue the rival merits of the two modes of operation; arguments there certainly are, as those who remember the virtual-circuit versus datagram debates in the 1970s will appreciate. The point being made here is that, apart from any technical issues, the very introduction of connectionless services gives rise to potential incompatibilities. If an independent choice of mode could be exercised at every layer, there would be so many varieties of OSI open system that the whole purpose of OSI would be vitiated.

The OSI community is well aware of this danger and, by placing limitations on the acceptable combinations of modes of operation in different layers, has severely limited the potential for this kind of incompatibility. Independent choices of mode may only be exercised either by the application-process (the 'user' of OSI) or in the Network Layer (where one finds the datagram versus virtual-circuit debate again in a new guise). However, it appears impossible to reduce this variety further. At the Network Layer, the proponents of neither mode will allow the other to be the mandatory 'conformance' mode; nor will the architects accept the only remaining means of removing the incompatibility ie the heresy of the transport relay.

Faced with this problem in the area of LAN/WAN interconnection, ECMA has adopted the ingenious stratagem of defining a new architecture for 'distributed end-systems', which uses OSI protocols but also permits relaying in the Transport Layer, in a device called a distributed end-system gateway. Whether this will affect OSI itself is as yet unknown.

In whatever direction one might ideally like to see ISO or CCITT progress, one must accept that there are practical and political limits to their power to enforce compatibility through international standardisation. The final step to complete compatibility must be taken by others. Ultimately it is the market that will decide: if the standardisers cannot agree the 'conformance' protocols, natural selection in the competitive market will make the choice. Thus to a large extent the responsibility rests on those who commission or buy open systems: if they want guaranteed compatibility with other open systems, they must insist upon it — and be prepared to pay for it. Some organisations are already trying to promote this natural selection and it is one of the objectives of the 'intercept strategies' which are discussed in the next section.

The adoption of OSI

The progress towards standards of the OSI services and protocols has not been even, some layers being very much ahead of the others. The most advanced are the standards for connection-oriented, namely:

- Network service
- Transport service and protocol
- Session service and protocol.

Each of these standards is already a DIS and all are expected to progress to IS this year or early in 1985. Unfortunately, this central block of well advanced standards is of little use on its own, because it requires a network protocol (as previously explained) to be mounted on top of real subnetworks; and even if that were present, the session service by itself is not intended to meet application process needs. Standards for the other layers are catching up, but one will have to wait until 1985 for a set of firm standards which will meet real application needs over a reasonable variety of communications technology.

The unevenness of this progress towards standards understandably generates some problems of its own. Not unnaturally, many users and suppliers wish to be poised to take early advantage of OSI, so there is a substantial user interest in the programme and a number of plans within companies to develop OSI systems. These are of course to be welcomed, but one consequence of this commercial pressure in the current, not-quite-stable standards environment is a tendency towards a proliferation of systems which are 'not quite OSI'.

One can readily see the motivation for a computer manufacturer, wishing to get into OSI, to build an architecture which includes the stable parts of OSI together with some private 'best guess' or 'minimal function' implementation of the not-so-stable parts. Unfortunately such an apparently sensible migration strategy could, especially if adopted with different technical choices by many companies, lead to incompatible systems on the market which would all no doubt be claimed to be OSI in some sense. Furthermore, once such systems were built and sold, it would not be easy to upgrade them immediately to 'real OSI', however well intentioned the migration strategy might be. Thus early efforts to adopt OSI in practice could result in the addition of even more incompatibilities between not-quite-OSI and OSI, above and beyond those already permitted by OSI standards.

'Migration', 'intercept' and 'harmonisation' strategies

A number of organisations have foreseen the dangers of incompatibility and have attempted to find ways of coordinating the adoption of OSI by substantial interested communities so that incompatibilities are minimised or eliminated. Such efforts are variously known as migration, intercept and harmonisation strategies. The significant ones in Europe are:

- The intercept strategy of the UK Department of Trade and Industry's FOCUS initiative
- Attempts by the Commission of the European Community to establish a common standards policy
- A product design strategy for the ESPRIT Information Exchange Systems (EIES) developed by 12 European manufacturers
- Architel, a strategy developed by the French PTT for the coordinated introduction of Telematics services with OSI (videotex, teletex, facsimile...)
- ECMA standards and technical reports (possibly).

The query against the last item is there because it is not clear to what extent ECMA members regard themselves as developing a coordinated migration plan to real OSI, as opposed to making contributions to the development of OSI or even (given their interest in standards beyond OSI) the extension of OSI to an architecture of wider scope. There may not be a uniform view on this question.

The stated purpose of the UK intercept strategy is:

> *To facilitate the rapid development and deployment of* [Information Technology] *systems to the advantage of both suppliers and users by bringing standards into use very quickly. It aims to support the development and use of IT products but not to direct or determine them — this is best left to the market-place. Thus it involves the identification of standards, their production, refinement, application and maintenance.... Thus the intercept strategy aims to identify and promote emerging standards without waiting for the completion of some of the lengthy formal standardisation procedures.*

In addition, clearly identified 'interim' recommendations will be made to fill gaps in the coverage of technically stable standards. This strategy includes a substantial element of promotion and encouragement of UK industry to reap the benefits of OSI. It endeavours to set out a consistent migration policy to OSI and in doing so consciously removes the scope for the incompatibilities discussed earlier. It requires both Class 0 and Class 2 transport protocol, for example, and it defines the field of application of connection-oriented and connectionless services. It is being developed by the

Department of Trade and Industry's Information Technology Standards Unit, which publishes its recommendations in technical guides, a substantial number of which will have been issued during the Spring of 1984.

The CEC's Working Group on Standards, which is responsible for advising on the standards policy to be adopted by the Commission, has welcomed this UK initiative and very much wishes to see a similar strategy applied to the European Community as a whole. Representatives of national initiatives, such as the ITSU, and of existing CEC projects, including ESPRIT, are participating in discussions aimed at securing the necessary international agreement to the technical content of such a strategy. The results of these discussions are being published in 'Guides to the common use of OSI standards' (COS) by the Working Group on Standards.

In parallel with these efforts, the 12 manufacturers collaborating on the ESPRIT Information Exchange Systems (EIES) have had to make technical choices on the use of OSI in the immediate implementations of the EIES. As their documents state, this is not a standards policy but a set of product design decisions that have had to be taken at a particular point in time. Inevitably, the existence of implementations of particular choices will tend to reduce freedom to manoeuvre. Nevertheless it is understood that the group of 12 do not regard their position on migration to OSI as frozen by the EIES decisions, and will continue to support efforts to reach a common standards policy.

Architel is in a rather different category in that it is specifically directed at the integration of CCITT Telematics services with OSI. It therefore restricts itself to a particular set of applications and, appropriately enough, is being developed by the French PTT, in consultation with government and industry. It is hoped that the OSI components of Architel can be accommodated within a wider European policy, but given the differences of scope it is not certain that this can be done.

The prospects for broad agreement are good. Inevitably, as the number of interested parties increases so too does the difficulty of agreeing unique solutions to particular problems. What can be expected is a common policy which commands very wide support but which still leaves just one or two difficult problems unresolved, with the effect that universal compatibility cannot be guaranteed. As observed earlier in this paper, it is then up to market forces to do the rest.

10: The interconnection of messaging systems

R H Willmott

Logica UK Ltd
London
UK

The current development of messaging services and technologies is of a divergent nature in terms of message transfer methods, message content, structure and management. In addition, current products offer increasingly complex facilities, whereas a basic requirement of any messaging service is the ability to pass messages between as many messaging service users as possible. This paper is an overview of this apparent conflict.

R H Willmott

Mr Willmott graduated with an Honours Degree in Computer Science at the Hatfield Polytechnic after an initial career in telephone switching systems with the UK Post Office. He then spent some time in Germany with the Gesselshaft für Mathematik und Datenverarbeitung where he worked principally in the area of data communications and text communications protocols. He is currently employed by Logica in the Information Systems Division and is active in the production of standards for digital networks and text communications.

The interconnection of messaging systems

Introduction

The current development of messaging services and technologies is of a divergent nature in terms of message transfer methods, message content, structure and management. In addition to this, current message communication products offer increasingly complex facilities. These factors increase the number of services and messaging terminals which are potentially isolated from each other due to incompatibilities. In contrast to this, a basic requirement of any messaging service is the ability to pass messages between as many messaging service users as possible. The reduction of the number of isolated systems is thus an important aim. This paper is an overview of this apparent conflict in terms of the current state of messaging technology and emerging standards designed to fulfil the needs of maximum interconnection.

The diversity of messaging systems

Messaging systems and terminal technology are characterised by diverse techniques together with ever increasing and complex services and facilities.

Message transfer

Telex, teletex and fax systems in the public domain operate a terminal-to-terminal message transfer system involving no intermediate storage. This is referred to as direct messaging.

Other public and private message transfer systems operate on a store-and-forward principle where messages are relayed over several communications links and intermediate switches to their destinations. This form of message transfer usually has multiple-addressing facilities.

More modern Computer-Based Message Systems (CBMSs) work on a store-and-retrieve basis by use of mailboxes associated with each subscriber to the service. CBMSs are normally accessed by terminals in an interactive mode.

There are thus three different methods of message transfer between messaging systems and terminals.

Message content

So far, the type of the message body itself has been of a homogeneous nature. Either character-coded text or coded images (facsimile) have been transmitted as individual messages. Current developments enable digitally encoded voice and graphics as well as combinations of all these message body types within a single message. Such developments have given rise to the need for conversions and compatibility agreements for complex messages such that terminals generating message bodies of differing types can meaningfully exchange messages.

Office communications system products are emerging which include all of the complexity of mixed body types, special message transfer methods, differing data communications protocols and which additionally support complex office procedures and filing. These are further dimensions that serve to exacerbate the interworking problems.

The potential terminal base is perhaps the most diverse component of all. Home computers, personal computers, videotex terminals, teletex terminals, telex terminals, word processors and to some limited extent fax terminals are all capable of exchanging messages given the appropriate conversion facilities and standards to achieve compatibility.

Naming and addressing

Given the diverse nature of the many systems that can be interconnected, the problems of naming and addressing seems insurmountable at the level of the individual system. This is because each system or service has its own methods and strategy of naming and addressing, and each has its own existing naming allocation within its strategy. Mapping between the multitude of naming/addressing mechanisms is impracticable. Names and addresses on similar systems would no longer be unambiguous.

The rationale for messaging standards

These incompatibilities prohibit the direct message exchange between many of these systems and terminals. This situation, together with the need for some form of interconnection, has resulted in the requirement of a single set of standards for message exchange onto which the majority of messaging systems and terminals may map, which has a suitably defined set of common services and which will also form a basis for future developments in interconnectable products and services. Such a standard for message system and terminal interconnection is the subject of much current effort in national and international standards committees and it can be expected that such standards will emerge shortly.

Emerging standards for messaging

In order that dissimilar messaging systems may exchange messages, compatibility must be ensured (or conversion gateways provided) in the following areas:

- Data communications services and protocols
- Application layer message transfer protocols and services
- Message body content
- Originator/recipient naming and addressing
- Management and directory services.

This applies to the transfer of documents as well as to simple 'informal' messages. Since the problems are similar, the standards should be identical to avoid a multiplicity of solutions.

This compatibility may be achieved by a set of standards applicable to each area listed above. These standards are being produced by CCITT, ECMA and ISO, who are all proceeding towards very similar standards except in the area of management and directory services.

Standards for data communications

Open Systems Interconnection (OSI) standards provide a set of services and protocols to allow different systems to exchange data in a consistent way. The services provided may be regarded as a set of tools which are of use to many communicating applications, of which messaging is one. The OSI standards are dealt with more fully elsewhere, and will not be treated further here.

Message transfer standards

As previously mentioned, three distinct modes of message transfer are possible, and might be considered candidates for standardisation:

- Store-and-retrieve message exchange
- Direct message transfer
- Store-and-forward message transfer.

Store-and-retrieve message exchange

Store-and-retrieve facilities (associated with mailboxes) do not imply any particular method of message transportation between systems. Indeed either direct (mailbox-terminal) or store-and-forward techniques may be utilised between systems in addition to the usual interactive access methods normally exercised by terminals. Strictly speaking, a store-and-retrieve system might be considered as a user of direct or store-and-forward message transportation services. For this reason, standards are not applicable for store-and-retrieve as a function of message transportation between messaging systems; store-and-forward and direct message interchange are sufficient. The connection of terminals to store-and-retrieve facilities, however, may well be suited to standardisation. Indeed, parts of the CCITT X.400 series (P3 protocol) are tending towards such a standard.

Direct message transfer

Direct message transfer is provided by the fax, telex and teletex services. A complete set of standards exists (F200, S70, S62) for these services as does a standard for the interconnection of telex to teletex. Direct messaging relies on the interconnection of two terminals using a single communications channel, hence both terminals must be active simultaneously. There are several consequences of this:

1 There is little or no scope for conversion of protocol or body content of the message other than by the intelligence at the communicating terminals. This implies a negotiating protocol. A standard text coding and protocol needs to be established across all participating terminals. This could lead to different classes of direct message communications systems which are isolated from each other.

2 Each terminal must be equipped with a communications channel and service access dedicated to message transfer as well as an internal process dedicated to message exchange tasks to guarantee a high enough availability to receive incoming messages. Clearly, personal computers, word processors and other small machines with limited intelligence and communications capability and which must be available for several purposes are not easily capable of participating in this type of message transfer without degrading the overall quality of service due to non-availability to receive incoming messages.

3 The above characteristics have required a very high degree of standardisation of the terminals used for messaging in terms of coding, bit rates, etc which has not allowed a multi-purpose use without an increase in cost. These are distinct disadvantages in the telex/teletex services and terminals.

Store-and-forward message transfer

The store-and-forward technique overcomes many of the disadvantages of direct message transfer. It allows the possibilities of multiple addressing with route and transmission optimisation and will readily support conversions between the requirements of dissimilar terminal types. This is because store-and-forward techniques allow a centralisation of the intelligence necessary to interconnect and convert between the differing characteristics of many diverse terminal populations.

Standards for this have been under preparation for some years and are due to become publicly available shortly. The next section considers these standards in greater detail.

Standards for store-and-forward message transfer

Standards for store-and-forward message transfer and a method of use have been under development by the CCITT, ECMA and ISO for some time. The standards are all based on original models of message transfer from IFIP WG6.5 on CBMSs.

Currently, the CCITT has virtually completed its draft recommendations (recommendations of the X.400 series) preparatory to ratification at the 1984 plenary. For this reason this paper will consider the proposals in detail. ECMA (Message Interchange Distributed Application — MIDA) and ISO (Message-Oriented Text Interchange Standard) are moving towards harmonisation with the CCITT X.400 recommendation. The main current difference is that whereas the CCITT considers only the interconnection of public administrations messaging systems, ECMA and ISO are also concerned with the interconnection of private messaging systems and connections between private and public administration systems. Hopefully all their efforts will result in a single coherent set of standards for general application.

The functional model of store-and-forward message transfer is shown in Figure 1. It consists of a central

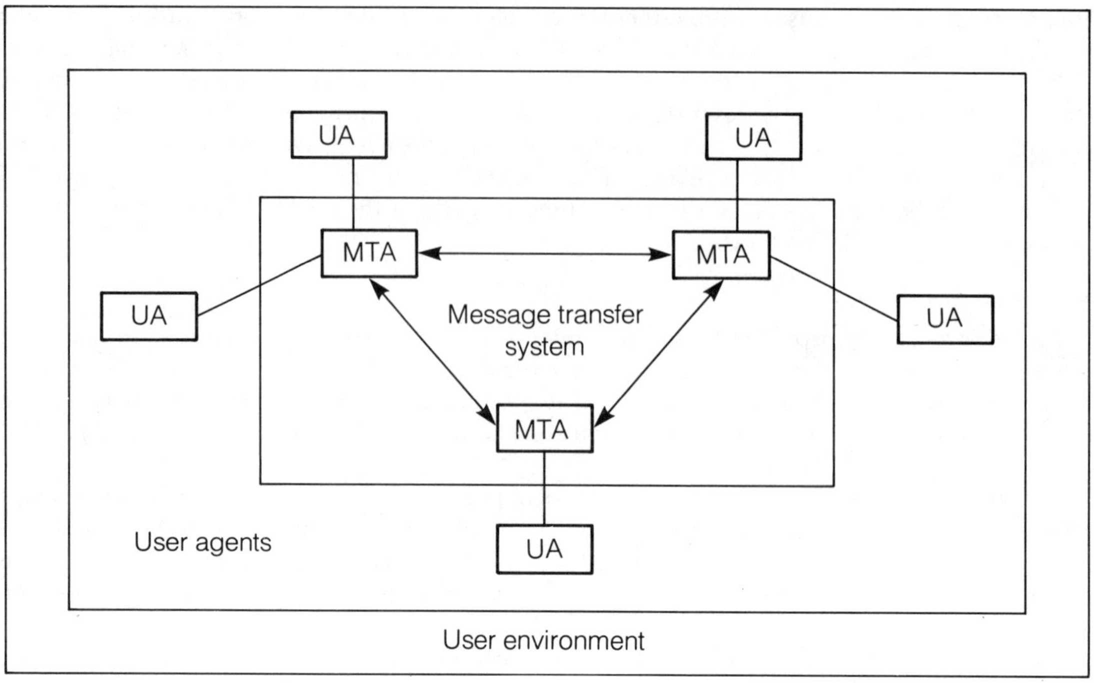

Figure 1: Functional model of store and forward messaging

set of Message Transfer Agents (MTAs) which cooperate rather like the switches of a message-switching network to relay messages between originator and recipients. Originators and recipients are represented to the message transfer subsystems by User Agents (UAs). UAs are uniquely identifiable objects which submit messages to and receive delivery of messages from the message transfer service on behalf of originating/recipient human or computer process user.

There is a corresponding two-layer model of this which is shown in Figure 2. The services and corresponding protocols are constructed using the same principles as the layering used in the OSI reference model. The two layers are considered to be resident in the application layer (7) of the OSI reference model, and hence have access to all of the facilities of the transport, session and presentation services for data communications. A layered model, including OSI and the messaging application is shown in Figure 3.

The Message Transfer Sublayer (MTSL)

The lower layer, called the Message Transfer Sublayer (MTSL) provides a store-and-forward message transfer service having the facilities indicated below:

1 Access management. Logon, Logoff facilities for UAs remote from their MTA.

2 Content type indication. An indication of the sort of message header within the message, hence indication of UA type to be used in decoding the message.

3 Converted indication. An indication that a conversion of the message body coding has/has not been made.

4 Delivery time stamp indication. An indication of when the MTA delivered the message to the recipient UA.

5 Message identification. A unique message identity generated by the message transfer layer on submission of each document.

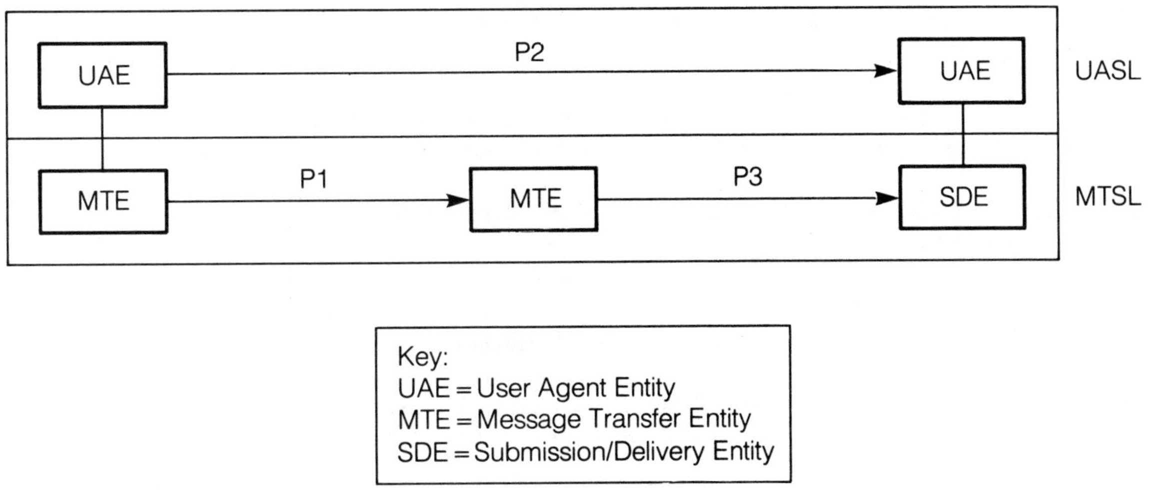

Figure 2: Layered model of the messaging application

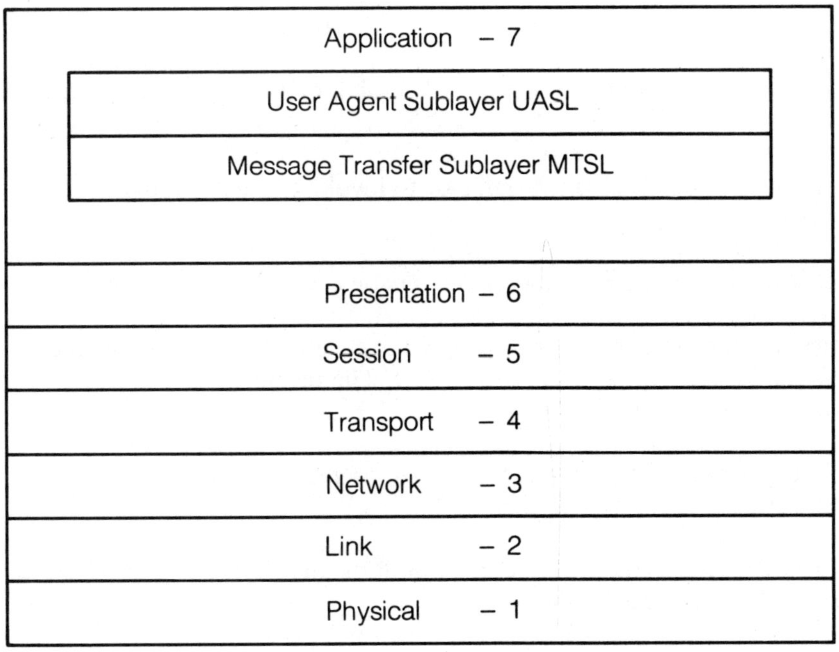

Figure 3: Relationship between OSI and messaging application

6 Non-delivery notification. An indication to the originator, that a message has not been delivered.

7 Original encoded information types indication. An indication of the message body coding before any body content conversion took place.

8 Registered encoded information types. An indication from a UA to its MTA which registered information types can be received (ie which body codes).

9 Submission time stamp indication. An indication of the time at which a message entered the first MTA from the originating UA.

10 Alternate recipient allowed. A means by which the originator can allow (or prohibit) the recipient UA to reroute a particular message to an (unspecified) alternate UA.

11 Deferred delivery and deferred delivery cancellation. A mechanism allowing an originator to specify a delivery time slot for a particular message.

12 Delivery notification. A notification generated by the terminal MTA when a message is passed to the recipient UA. This notification is passed back to the message originator.

13 Disclosure of other recipients. Messages are associated with a list of intended recipients. This service allows disclosure (or non-disclosure) of the recipients' list in each individual's received message.

14 Grade of delivery selection. Allows a message originator to specify whether a message should be forwarded normally or urgently.

15 Multidestination delivery. A single message to multiple recipients will be submitted as a single message content and an associated recipients' list. Copies of the single message content will be made at appropriate MTAs along the route where different copies must take different routes (due to optimisation etc).

16 Prevention of non-delivery notification. A means by which non-delivery notifications may be suppressed.

17 Return of contents. Allows the originator to request the return of message contents if delivery is not possible.

18 Conversion prohibition. A means by which the originator can inhibit the body content conversion mechanism.

19 Explicit conversion. Allows a message originator to invoke a specific conversion of body content for a particular message.

20 Implicit conversion. Allows specification of conversion for all messages sent from a given originator.

21 Probe. A service which enables the submission and trial delivery of a 'dummy' message to establish the deliverability of a message of given size, content etc. The results of such a probe will be a non-delivery or delivery notification.

22 Alternate recipient assignment. This allows a UA to redirect undeliverable incoming messages to another (specified) UA for a given time period.

23 Hold for delivery. This allows the UA to detach from the messaging service MTA for a period of time. Messages arriving will be stored at the MTA.

Two protocols are associated with the MTSL.

P1 is a protocol suited to the interconnection of MTAs (message switches, gateways, CBMSs and certain terminals to the system). It is similar to a file transfer protocol but has a multiple address facility and other messaging-specific characteristics.

P3 is a protocol allowing single UAs remote from their MTA access to message transfer layer services. It is an interactive protocol and allows the UA and terminal to be disconnected from the MTA. P3 represents a boundary of responsibility for messages submitted to or delivered from an administration and a boundary of functionality (ie the P3 protocol does not allow access to all of the MTSL services). The P3 protocol does, however, have some elements of store-and-retrieve services allowing the terminal to instruct the MTA to hold messages for delivery.

The User Agent Sublayer (UASL)

The UASL protocols operate between peer UA entities and form the association between corresponding parties for any given message. An analogy here is that while the message transfer sublayer and protocol represent a normal postal envelope, the UASL and protocols represent the message heading.

UA services include all of the services of the message transfer layer plus a set of cooperating UA actions and a set of information-conveying indications.

1 Cooperating UA actions:
- Blind copy recipient indication: originators of multiple addressed messages may require that certain recipients are not aware of the identities of other intended recipients; this service element instructs the originator's UA to construct multiple copies of the messages containing appropriate address lists
- Non-receipt notification: a request for the originator to be informed if a message could not be delivered to the specified recipient
- Receipt notification: a request that the recipient UA (or person) should indicate that the message has been received
- Autoforward indication: allows the recipient of an incoming message to determine that a message has been autoforwarded from another UA.

2 UA information conveying indications: the following are standard fields which are used for information only:
- Originator indication
- Authorising users' indication
- Primary and copy recipients' indication
- Expiry date indication
- Cross-referencing indication
- Importance indication
- Obsoleting indication
- Sensitivity indication
- Subject indication
- Replying IP-message indication
- Reply request indication
- Forwarded IP-message indication
- Body part encryption indication
- Multi-part body.

Some of these elements reflect the services of the MTSL and are directives to that layer. Some elements are directives to the recipient UA which may be interpreted and acted upon automatically (such as autoforward, recipient notification). The other elements are information to the recipient regarding the message.

The UA services may be used to obtain a quality of service suited to individual messages. An informal message (second-class post) might not invoke the non-delivery diagnostic nor request for notification of receipt. An important document requiring a high quality of service (registered post?) will invoke these services to ensure delivery and note receipt.

This set of standard services and protocols for store-and-forward message transfer (the X.400 draft recommendations) provide a flexible means of interconnecting many diverse messaging systems, CBMSs, message-switching systems and, in particular, standard ways of connecting the telex/teletex community to the messaging community. CBMSs and gateways will be developed to allow other isolated communities such as videotex, word processors and personal computers to access the messaging community.

This all goes to fulfil the basic need to enable the physical movement of messages between diverse systems and terminals.

Naming and addressing

In the X.400 draft recommendations the problems of naming and addressing referred to earlier in an environment containing many different systems has been overcome by overlaying common (standard) name *forms* to identify uniquely all UAs in the messaging community. In parallel with this, a naming domain hierarchy is used to administer names within a specific section (domain) of the total messaging community independently of any other section.

A valid name is an identifier that will cause a directory to generate a unique address of the specified UA. The address gives the location of the UA. An address can be either completely or partially translated into a route through the message transfer service using routing tables.

The current name forms defined by CCITT are given in Figure 4. The names chosen also happen to be addresses which are automatically translatable into routes by routing tables. To avoid confusion here, the term 'address390 will be used for CCITT name/address terms. It may be expected that other name forms which are not also directly usable as addresses (and will need a more complex directory translation for the address) will be defined.

The addresses are used in the headings of messages (P1, P2 protocols) in one of the standard forms. This is translated into domain specific names (non-standard form) as a message enters the domain. The standard address may be reinterpreted as the message transits different domains.

In practical terms, each group of users having a specific addressing convention and corresponding address space will become a domain. Associated with each domain will be a set of translations from the address in standard form to the address in domain internal form and hence into routing information suitable to that domain. It is not necessary that each translation makes use of the whole address. Only sufficient parts of the address to ensure an unambiguous translation to routing information is necessary.

Message content compatibility

The message transfer standards previously described are capable of moving a message between any two terminals irrespective of the type of message content produced by the originator or the capability of correctly rendering the message content type at the recipient terminal. Several dimensions to the extent and resultant potential incompatibility exist:

1	Country name	
	Administration domain name	
•	Private domain name	
•	Personal name	
•	Organisation name	
•	Organisation unit name	
•	Domain-defined attributes	
2	Country name	Numeric keyboards
	Administration domain name	
	UA numeric identifier	
•	Domain	
3	Country name	Telex
	Administration domain name	
	X.121 address	
•	Domain-defined attributes	
4	Telematic address, (ie X.121)	Telematic terminals
•	Telematic terminal identifier	

• = Optional.

Figure 4: CCITT name/address forms

1 Message body content type. The information may be character coded text (IA5; teletex; telex...); fax (Groups 1, 2, 3...); graphic information; or digitally encoded voice (various codings)

2 Logical structure of information content. Messages are structured into logically interconnected hierarchies (documents, sections, subsections, paragraphs, etc) which reflect the logical dependencies of information items.

3 Layout structure. Messages may contain elements in which the exact layout of the content is of importance (ie diagrams, tables). These layouts can be disrupted by incompatible line widths, etc.

Possible solutions depend on the scale of the problem and nature of messages produced by a particular terminal type. Messages of a monolithic body type (ie consisting of only a single body type) may be converted into another (monolithic) type while the message is in transit through the message transfer service. Thus, an IA5 coded message can be converted for presentation on a teletex (or even fax) terminal by special processing. This scheme is ideally suited to the majority of less intelligent terminals such as teletex, fax, teletypes, etc which produce text of a limited and simple type. This method of conversion has been adopted within the CCITT message transfer service and a corresponding standard specifying the conversion rules has been produced (X.408). It may be implemented in the central logic of the MTAs.

Such a solution is not readily applicable in the case of more complex terminals which might be capable of combining character coded text, fax and other types into a single complex message. This is a case of associating the message content with a specification of message body content type, logical and layout descriptions. These descriptions are forwarded with the message content to enable decoding of the message content at the recipient's terminal. This clearly requires a high degree of intelligence at the recipient's terminal and is therefore unsuited to the simpler terminals.

Standards for this type of message content description are currently being developed by CCITT (T.73); ECMA (office document interchange format) and ISO (office document architectures).

Management and administrative aspects

The technical standards presented so far will eventually enable the wide-scale interconnection of very diverse terminals and systems for the purpose of message exchange. Other technical standards and solutions to management and administrative problems remain to be produced before a viable integrated service can achieve its full potential.

The directory functions

Finding the machine interpretable name or address of potential message recipients is traditionally aided by a 'directory enquiry' service as associated with the telephone service. The equivalent service for messaging will be a more complex task because a given recipient may:

- Have more than one terminal/UA
- Have more than one terminal type
- Belong to more than one service
- Have invoked autoforwarding options
- Have several possible types of name (CCITT name forms).

Directories may therefore contain multiple entries for each potential recipient. There is a clear need for user-friendly directories having user interfaces conforming to international standards. In parallel with this, there is also a need for standards enabling the update of the 'directory enquiry' databases (which might be replicated and partitioned) in a consistent and synchronised manner with the message transfer system routing directories (also a replicated, partitioned distributed database).

Management and administration functions

Whereas the CCITT, ECMA and ISO have provided structures and name forms for use in identifying originators and recipients, naming authorities are needed to allocate the actual names to originators and recipients of the message. Currently, no such administration exists on a national nor international basis. Foundation of such a thing is a prerequisite to successful communications. It may well be that this administration and management will be arranged internally between the PTTs and publicly available value-added network operators, but this is likely to preclude the *direct* interconnection of private (in-house) systems not wishing to use the services of VANs or PTTs.

Man/machine interface

Diversity and complexity of terminals connected to different systems, each exhibiting widely divergent local characteristics, will cause great difficulty to non-technical users who should be capable of operating from different terminals. The telephone is easy to use because of its simplicity, but messaging terminals are far more complicated. A standard interface will become urgent in the near future to avoid confusion and disillusionment among the end users of these interconnected systems.

Conclusions

Technical standards enabling meaningful message exchange between many diverse types of system and terminal will be available shortly from CCITT and ECMA and later from ISO. It seems that these standards will be compatible at least at the message transfer level and hopefully at the document content levels. Pilot implementation and services can already be considered to verify the standards.

Technical standards are lacking in the areas of man-machine interface (for messaging) and promoting user friendliness.

Administrative and management functions must be urgently defined and established to allow a wide scale and flexible use of the fruits of the technical standards.

Invited Paper references

BUR1
Davies E
'Telecommunications, a technology for change'
HMSO (1983)

BUR2
Beesley M E
'Liberalisation of the use of British telecommunications network'
HMSO (1981)

BUR3
Financial Times p 12 (27 Sep 1982)

BUR4
Financial Times p 11 (21 Oct 1982)

BUR5
Financial Times p 5 (15 Nov 1982)

BUR6
Computer Weekly p 7 (4 Dec 1982)

BUR7
Electronics Weekly p 3 (3 Nov 1982)

BUR8
Winsbury R (editor)
'Viewdata in action. A comparative study of Prestel'
McGraw-Hill (1981)

BUR9
'Cable systems'
Rep by the Information Technology Advisory Panel
HMSO (1982)

BUR10
Rep of the Inquiry into Cable Expansion and Broadcasting Policy
HMSO (1982)

BUR11
Institute of Information Scientists
Submission to the Hunt Committee of Inquiry on cable broadcasting (1982)

BUR12
Wolfe R
'Videotex, the new telephone information services'
Heyden p 152 (1980)

BUR13
Fox B
'Will the Dutch cable Britain?'
New Scientist pp 24-25 (29 Mar 1984)

BUR14
Financial Times p 3 (28 Mar 1984)

BUR15
'The cabling of Europe. In the shadow of the satellite'
The Economist pp 49-56 (13 Aug 1983)

BUR16
'Information processing. Continental telephone. Taking on the giant in telecommunications'
Business Week Intl edition pp 38-42 (9 Feb 1981)

BUR17
Sackman H and Nie N (editors)
'The information utility and social choice'
AFIPS Press US (1970)

BUR18
Gillespie G
'Public access cable television in the United States

and Canada'
Praeger Publishers (1975)

BUR19
Jacobson R E
'Municipal control of cable communications'
Praeger Publishers (1977)

BUR20
'The plan for an information society. A national goal towards the year 2000'
Ekistics pp 175-182 (Sep 1974)

BUR21
Financial Times Special Supplement
p viii (12 Dec 1983)

BUR22
Electronics Weekly p 12 (4 Mar 1984)

BUR23
Financial Times p 13 (5 Sep 1983)

PRI1
International Standards Organisation
'Information Processing Systems — Open Systems Interconnection — Basic Reference Model'
International Standard 7498

PRI2
National Bureau of Standards
Data Encryption Standard. Federal Information Processing Standard 46 (15 Jan 1977)

PRI3
American National Standards Institute
'Data Encryption Algorithm'
ANSI Standard X3.92-1981

PRI4
National Bureau of Standards
'DES modes of operation'
Federal Information Processing Standard Publication 81 (2 Dec 1980)

PRI5
American National Standards Institute
'Modes of operation for the Data Encryption Algorithm'
ANSI Standard X3.106-1983

PRI6
National Communications Systems
'Interoperability and security requirements for use of the Data Encryption Standard in the Physical Layer of data communications'
Federal Standard 1026 (3 Aug 1983)

PRI7
American National Standards Institute
'American National Standard for Information Systems — Data Link Encryption'
ANSI Standard X3.105-1983 (Aug 1983)

PRI8
Price W L
'Key management for data encipherment'
Proc IFIP/Sec'83 Stockholm pp 205-215 (May 1983)

STO1
Rosenthal R M
'Network access techniques — a review'
Proc AFIPS NCC 45 pp 495-500 (1976)

STO2
Metcalfe R M and Boggs D R
'Ethernet: distributed packet switching for local computer networks'
Communications of the ACM vol 19 no 7 pp 395-404 (1976)

STO3
'The Ethernet: a local area network; Data Link Layer and Physical Layer specifications (Version 2.0)'
DEC/Intel/Xerox (1981)

Analysis

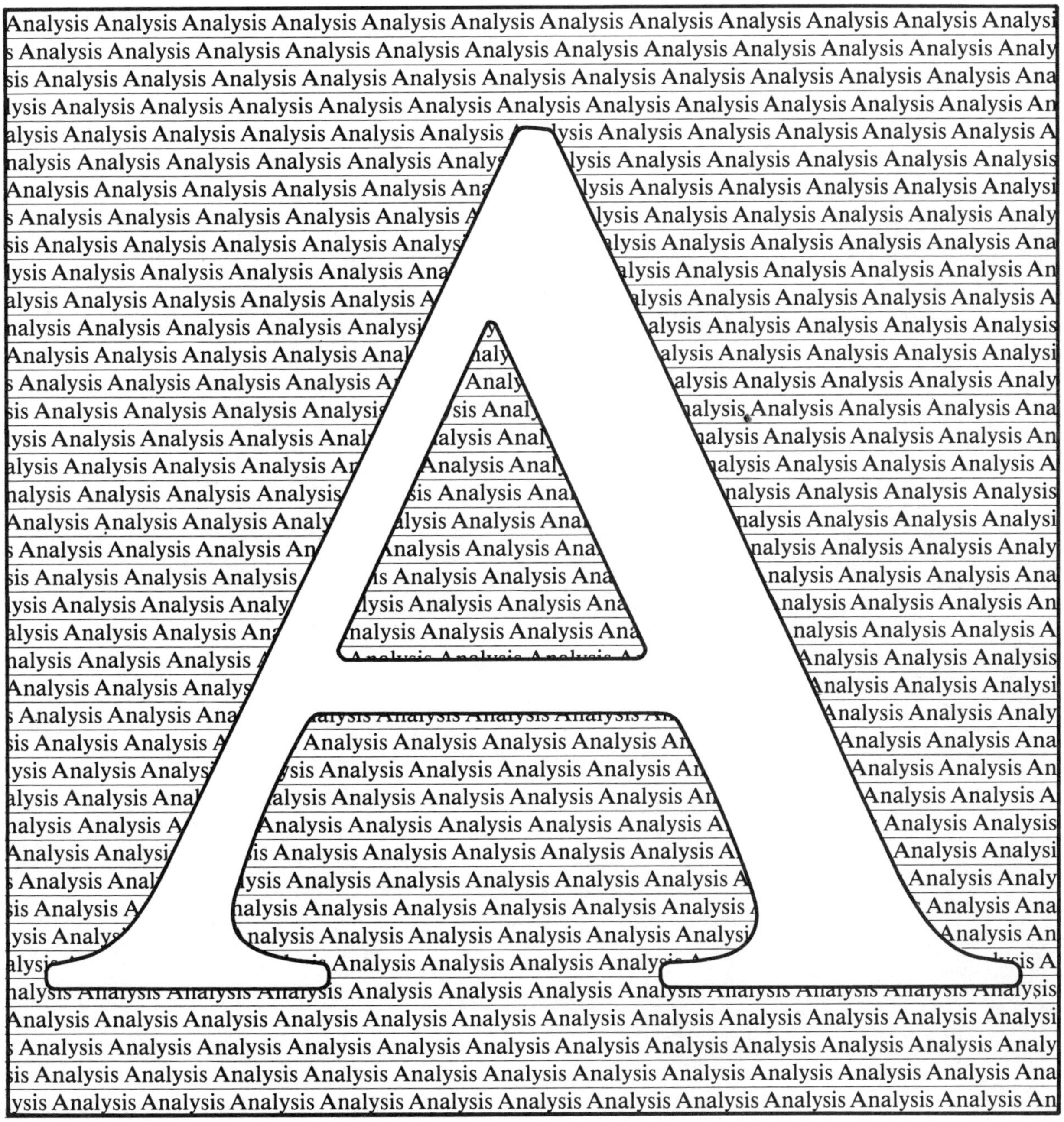

1: Requirements

Many organisations are equipped with a variety of systems, each installed for different purposes. These may include data processing applications, word processing, telex and message systems, telephone, facsimile, video, personal computers etc. In many cases, these systems are the responsibility of different people in the organisation. The systems perform isolated functions, but often the information they carry or contain could well be shared — to the benefit of users and the organisation. Integration of their operation as well as of the communication resources they use, represents an opportunity for increased efficiency.

Requirements

The status quo

Office automation, or perhaps more properly office mechanisation, is by no means the ultimate goal of integration in the corporate workplace. However, the topic may serve as an interesting example of an equivocal situation.

Meade (001):
Responsibility for this is often unclear. A recent survey of 60 major American companies asked who was in charge of office automation. The results are illustrated in Figure (1)1. The results of the survey would indicate that there does not seem to be a 'right' place for office automation responsibility and that, over a period of two years, there was little change in this equivocal situation. The initial implementation of word processing equipment did not clearly fall into the hands of the DP department. However, the exploitation of word processing seems to require systems expertise. Local area networks and data communications are usually seen as developments from the DP side, often including experts from both systems and operations.

Several interesting threads can be picked out from this example. Not only is the responsibility for the application of technological developments scattered about within organisations, but the possibilities for harmonisation and/or integration of systems is prejudiced by the division of responsibilities (as exemplified by the classifications 'DP types', 'administration types' and 'communication types' etc). A typical organisation will employ a variety of technologies supporting a variety of functions (found desirable or necessary to the business of that organisation) which may include the following:

- Electronic data processing (in mainframe and minicomputers)
- Word processing (stand-alone or shared logic)
- Personal computing (micro on the desk or 'timesharing'
- Message systems (eg telex)
- External information systems (eg Prestel or Reuters)
- Telephone (both PABX and public network)
- Other (eg videoconferencing, facsimile etc).

Burren:
We thus have the position where there will be an enormous increase in the number of computer-driven devices over the next few years. Users of these devices will wish to exchange information in the form of text, graphics, data, images and voice. The basic technologies for transporting very large volumes of information are currently available and have enormous growth potential.

Within an organisation these systems and their users may be geographically distributed. This gives rise to the now quite common use of telecommunications between the scattered elements. However, it is not

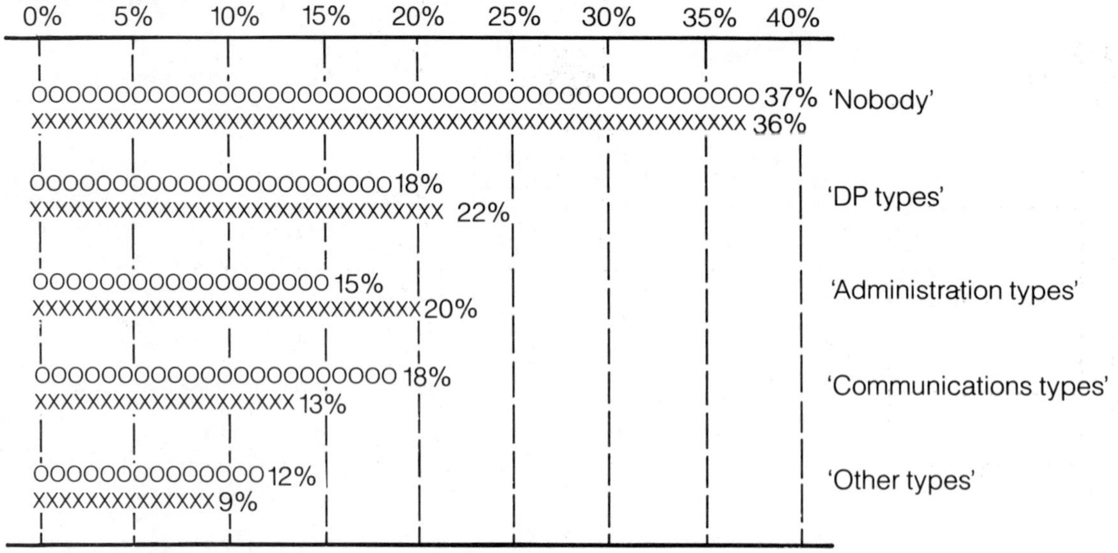

Key: OOOOOOO = 1980 survey findings (based on 33 companies)

XXXXXXXXX = 1982 survey findings (based on 60 companies)

Note:
Use of terms such as 'DP type' or 'Administration type' will offend serious 'statistical types', but they do not have to deal with the euphemistic job titles surveyed, for example, 'DP types' include the MIS director, vice president, information systems, data processing manager etc.

Figure (1)1: Responsibility for office automation

uncommon to find that each application subsumes its own telecommunication network since there may have been no *prima facie* case for combining them (see the vertical dimension in Figure (1)2).

Where telecommunication resources are shared, it is usually on the basis that common routes provide an opportunity to reduce total expenditure with the public telephone carrier. Examples are the shared use of leased telephone lines for voice and data, or the use of a PABX, perhaps for some dialled-up data calls.

Typically, each of the systems (horizontal dimension in Figure (1)2) 'belongs' to a different part of the organisation. The main computing facilities are run by an EDP department, for example; the word processing is provided on a group or divisional basis, while telephone and telex is the responsibility of the local administrative services. Individuals may or may not be permitted to buy the personal computer of their choice.

Opportunities for integration

There are good reasons for considering 'lateral integration' besides the sharing of telecommunication resources. These relate to:

1 The ability to share, on a corporate basis, data and information which resides or is processed in independent systems.

2 Making a wider selection of corporate resources available to individuals who could profitably use them.

Two simple examples will serve to illustrate these considerations. Firstly, text processing and message facilities are increasingly to be found in many systems.

> *Willmott:*
> *The current development of messaging services and technologies is of a divergent nature in terms of message transfer methods, message content, structure and management. In addition to this,*

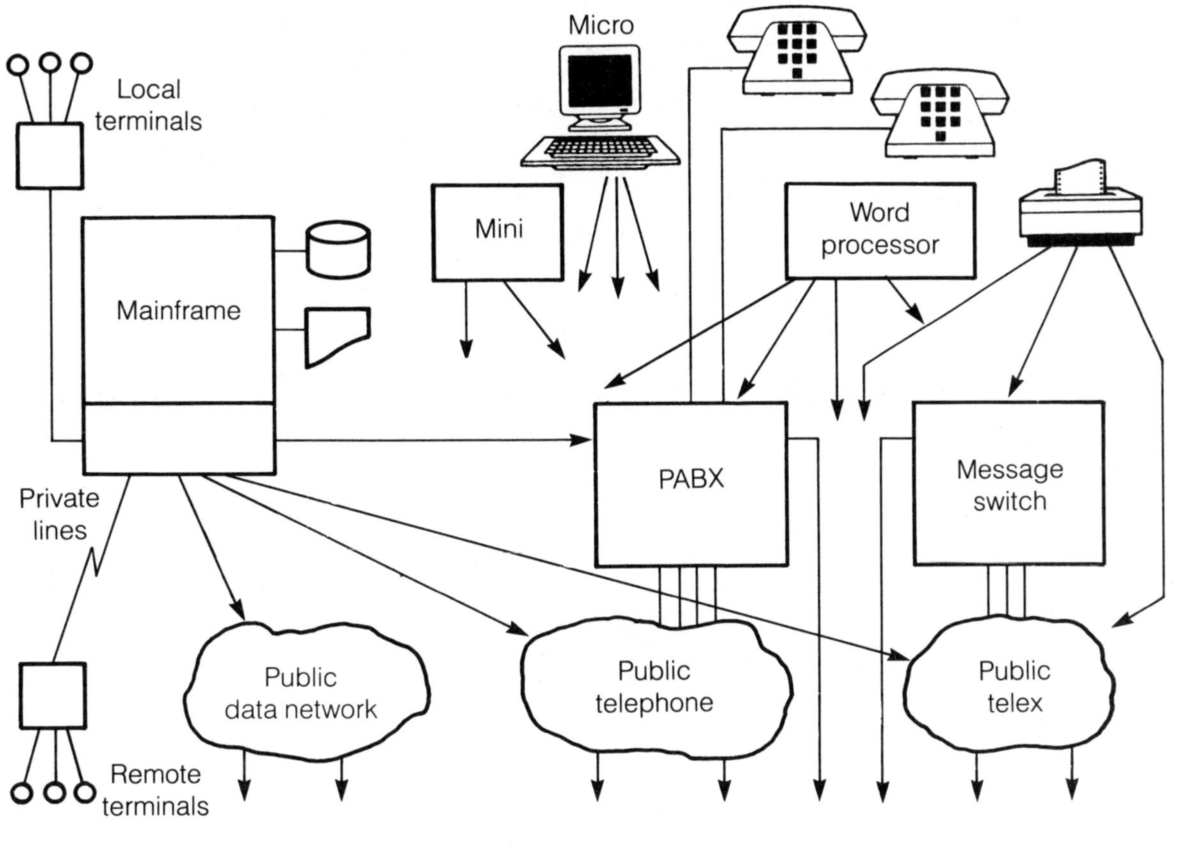

Figure (1)2: Current situation

> *current message communication products offer increasingly complex facilities. These factors increase the number of services and messaging terminals which are potentially isolated from each other due to incompatibilities. In contrast to this, a basic requirement of any messaging service is the ability to pass messages between as many messaging service users as possible.*

In EDP systems text editors are provided primarily for the purpose of creating and modifying program source text, but in many EDP networks such editors are used to create messages which can be inserted into a user file or 'mailbox' from which they can be retrieved by the nominated recipient *(002)*.

Contrast this situation with the preparation of text in a word processor which may also have data transmission or 'electronic mail' facilities. One immediately obvious difference is the nature of the user interface, one being designed for programmers, the other for secretaries and typists. Another difference will be the way that such messages are distributed — as exemplified by the Wang Mailway system *(003)*.

Compare the above systems with the message facilities of telex and the telegraph message-switching services provided in the public domain and it will be noticed that, despite the differences, a considerable overlap exists. In a recent study jointly funded by the government's Electronic and Aeronautics Requirements Board and the National Computing Centre it was concluded that mailbox systems will have a dramatic impact on productivity and effectiveness (see Figure (1)3).

> *Wilson (004):*
> *The potential impact of mailbox systems on productivity was identified after investigating some of the most advanced user organisations. It is clear that when a mailbox system has permeated throughout an organisation, a new corporate culture starts to emerge.*
>
> *This is characterised by less concern with processes and more concern with getting things done; delegation becomes more common and employees are more flexible and responsive.*

In fact all three of these types of message handling may well exist in the same organisation but may or may not be capable of interworking. Despite the fact that some word processors are equipped with telex

Organisation	System	No. users reached
Istel	Comet	400
Zynar	Zynar/Nestar network	120
Plessey	Internal Plessey system	150
Datapoint	Datapoint Electronic Mail Service	200
Texas Instruments	Texas Instruments Inc. MSG System	1000
Mailnet	Mailnet	150
Computer Sciences	CSC Notice	1000
Geisco	Geisco's Quik-Com System	200
Milepost Business Systems	British Telecom NE internal systems	50
	Janet Arpanet	{ 300
ICL	ICL internal systems	500
Bell Northern Research	Bell Northern Research Cocos	4950
GMD	Komex from GMD (West Germany)	370
Rank Xerox	Xerox Internet	5000
	STSC APL plus SIG Icebox	200
DEC	Digital internal (all-in-one)	100
	Usenet	5000
NPL	NPL Scrapbook & Edit Systems	180
Hewlett-Packard	HP's internal HP Desk Manager network	2000
New Jersey Institute of Technology	EIES at NJIT	200
I P Sharp Associates	I P Sharp Associates Mailbox	7600
Prestel	Prestel Mailbox	31 000
Stockholm University Computing Centre - QZ	COM	2000
NCC	NCC's Zynar network	20
	Blend	40
		62 730

Figure (1)3: Organisations and systems taking part in the international dissemination of mailbox recommendations on 17 January 1984

transmission capabilities, true interworking in a unified manner has yet to be provided on a widescale commercial basis — and this is just for the simple application of delivering text messages. Nonetheless, it is not difficult to imagine the potential benefits of being able to interwork between the different machines used within an organisation; indeed it is possible, when one considers the existence of networks like CHAPS which carry monetary transactions between the various computer systems of the major clearing banks *(005)*.

A second example relates to the user interface, particularly the user's terminal. All too frequently a user is provided with a terminal, or access to a terminal, which is essentially dedicated to a particular computer system (or even a particular application). If the user wishes to access a variety of different systems he or she frequently has to make use of a different terminal for each. A better solution would be to use the same (familiar) terminal for a wide variety of services.

> *Willmott:*
> *The potential terminal base is perhaps the most diverse component of all. Home computers, personal computers, videotex terminals, teletex terminals, telex terminals, word processors and to some limited extent fax terminals are all capable of exchanging messages given the appropriate conversion facilities and standards to achieve compatibility.... Diversity and complexity of terminals connected to different systems, each exhibiting widely divergent local characteristics, will cause great difficulty to non-technical users who should be capable of operating from different terminals. The telephone is easy to use because of its simplicity, but messaging terminals are far more complicated. A standard interface will become urgent in the near future to avoid confusion and disillusionment among the end users of these interconnected systems.*

It matters not to the user how this is achieved — whether it is by the use of intelligent terminals which can emulate different appearances according to the service being accessed, or by the interlinking of the service-providing computers, or by the adoption of common standards for terminals and message formats.

As the adoption of personal computers on the desk increases, individuals observe that connection to a telecommunication network opens up the possibility to use the device in more interesting and powerful modes.

The above examples reveal that unless the requirements for integration exist, the opportunities may not be recognised or adopted. This is due to two factors:

1 Lack of foresight and planning.
2 Cost considerations.

In the end there must be a recognition of the benefits and needs for information flow (an understanding of the technical possibilities will have a catalytic effect on this).

Corporate structure and data flow

Sizes and structures of organisations vary, but generally it will be found that most strategic information flows up and down the hierarchy. Additionally, many organisations can be divided into the following two levels of hierarchy:

1 A central corporate part.
2 A set of otherwise independent operating groups or companies.

This is shown diagrammatically in Figure (1)4.

The central corporate part of the organisation will be concerned with a number of factors, including the following:

1 Performance and profitability of the various groups.
2 Investment potential and requirements.

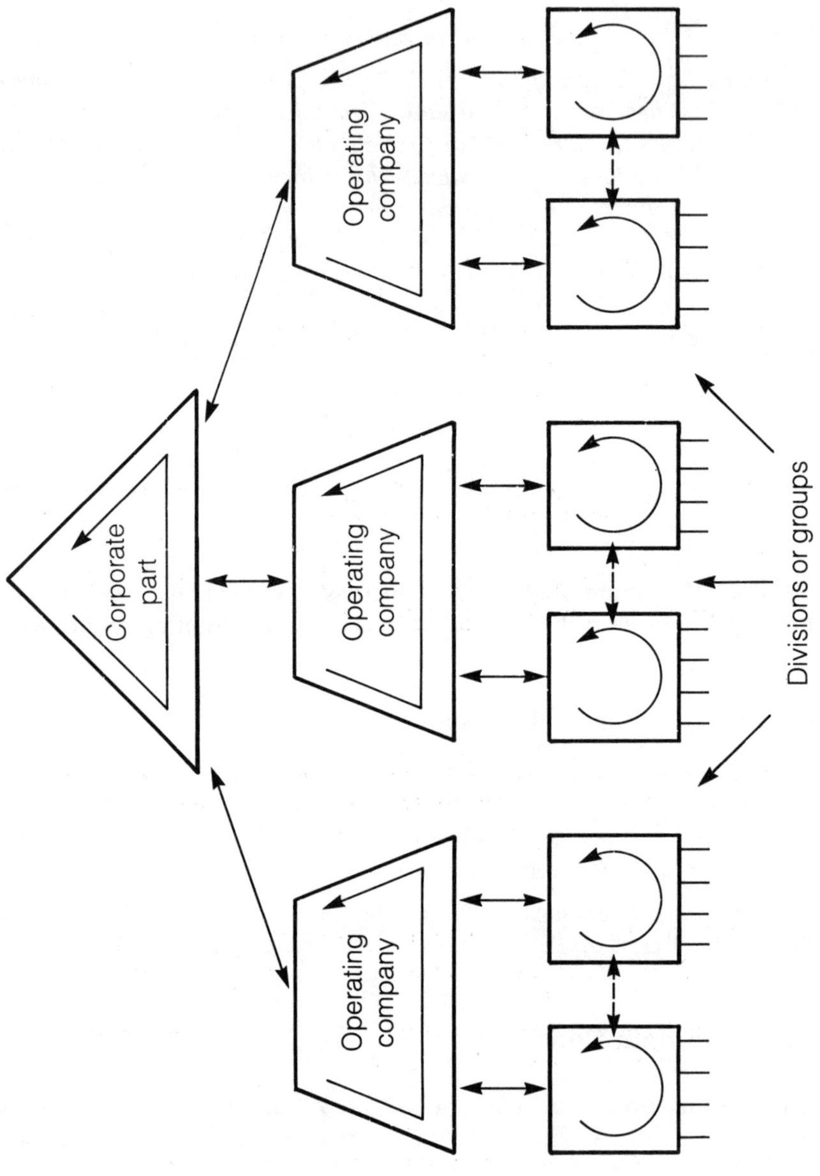

Figure (1)4: Corporate structure and information flow

3 Cash management at the corporate level.
4 Need for rationalisation and reorganisation as seen from the global viewpoint.

The operating groups and their various subgroups are usually required to function as individual profit centres. This means that they are concerned with such factors as:

- Cash flow
- Maintaining or improving efficiency and profitability
- Communicating with customers and suppliers
- Controlling staffing levels
- Agreeing and meeting targets
- Day-to-day operations.

> *Stokes:*
> *Frequently, users within the hospital are able to purchase computers with, for example, research funds and then wish them to be connected to other systems (in particular to the computer containing the patient master index). There is often little, if any, control over the computers chosen and so a highly flexible approach is needed; this can be provided by adopting the distributed strategy.*

Contrast then, the natural pressures on these two levels: the corporate part will need to exchange similar kinds of information with the various operating groups and may therefore be interested in a standard set of procedures, while the operating groups are interested in the local optimisation of their data and information processing and not in the problems or special requirements of other groups, except insofar as their solutions could prove to be useful. Indeed, since the cost of telecommunications and computer systems is usually borne by the operating groups, they will tend to install systems that meet local rather than global criteria. This leads to a number of local suboptimised solutions, rather than a global optimisation.

Clearly the central or corporate view would tend to conflict with these local views, but since the corporate organisation is usually not funding the systems, it cannot expect the local profit centres to be sympathetic to any imposed solution which could cost more for some of them. On the other hand, the *non-coordinated* approach to such systems results in procedural differences which tend to build walls around the individual groups. This in turn militates against the flexibility with which the corporate planners can group or merge operations.

> *Stokes:*
> *However, a number of problems were evident, arising from the (semi-)autonomy of the 'owners' of these applications. First, the data needed within the hospital was distributed over a number of processors not centrally controlled. Thus it would be possible for data to be unavailable at critical times. Also, the method of accessing data would vary from machine to machine (or, from the user's point of view, from application to application).*

Furthermore, the specification and provision of some teleprocessing equipment may reside partly with individual groups and partly with the corporate part (eg a central DP mainframe).

It can be readily seen, therefore, that corporate requirements tend to be different from those of operating groups; the policy must recognise and deal with this situation.

> *Stokes:*
> *In order to overcome these problems, the strategic plan proposed that there should be a particular system on the network providing any services which needed to be shared among the users of the network; this was called the shared services system.*

Technology is, of course, neutral in this — it can help or it can hinder, depending on whether or not its use has been properly planned.

Corporate standards

There are two types of standardisation *de jure* and *de facto*. In one sense it does not matter which standards are adopted, the important decision is what to standardise. The easiest things to standardise tend to be those over which little choice may be exercised. In the past this has usually been the telecommunication aspects (mainly due to the monopolistic position of the major supplier, the PTT or regulated carrier). In recent years, however, this situation has changed with a greater choice of network services being offered. Examples are the new public data networks, the introduction of digital transmission services, and, in the UK, the deregulation of British Telecom, resulting in an increased choice of consumer equipment which includes the following:

- Terminals
- Statistical multiplexors
- Packet switches
- Local area networks
- Wide area networks
- Network controllers and testers
- Protocol converters/gateways
- Value-added services.

Despite these changes, the telecommunication aspects still present a reasonably constrained set of *de jure* standards from which to choose. More problems present themselves in the choice of computer systems and their terminals.

If we update an old adage we might say that you can bring your computers to a network, but you cannot make them converse. This situation arises because of the wealth of choice of such devices and lack of *de jure* standards. *De facto* 'industry' standards tend to rely on particular manufacturer's products, but they only provide a limited solution. For example, many computer manufacturers provide communication products which use the IBM 2780/3780 protocols. This does indeed allow blocks of data to be passed between different computer devices. Unfortunately the problems still remain of deciding how the data shall be interpreted — if this is application dependent then there is no general solution for interworking between any pair of devices found in the market-place. Add to this the fact that the above-mentioned IBM devices are not supported in IBM's new SNA systems and it can be seen that such 'standards' can only be regarded as *ad hoc* and partial solutions.

On what then should one standardise? Should it be the equipment used, the software packages or the interdepartmental procedures? The problem here is that standardisation is often thought to reside in the physical devices used. This is usually only true if all products are bought from a single manufacturer (and not always then — how many computer manufacturers also make a compatible PABX?). Of course one can purchase from the 'plug compatible' manufacturers, but the result is the same — increasing dependence on a single-manufacturer product line (see Figure (1)5).

Common formats and procedures would appear to offer an appropriate avenue, but since many of these are vested in the manufacturer's software products that go hand-in-hand with the hardware, one can rapidly find oneself facing the same problem.

> *Gurugé (014):*
> *Some current IBM distributed data processing products that can be configured to contain I/O components that conform to the SNA specification are as follows:*
> - *Teleprocessing access methods:*
>   - *— ACF/VTAM*
>   - *— ACF/TCAM*
>   - *— ACF/VTAME*
> - *Application subsystems:*
>   - *— CICS/VS*
>   - *— IMS/VS*
>   - *— DPPX*
>   - *— DPCX*
> - *Communications controller, control programs:*
>   - *— ACF/NCP*
> - *Terminals or devices:*
>   - *— 3270 Information Display System*
>   - *— 3730 Distributed Office Communication System*
>   - *— 3600 Finance Communication System*
>   - *— 8100 Information System*
>   - *— 6670 Information Distributor*
>   - *— SERIES/1 minicomputer etc.*

Excellent though such products as IMS, CODASYL and DISOS may be, they are not available on every computer. In any case products are only solutions to problems they fit. Since every organisation is

- IBM-compatible
  - 2780/3780
  - Hasp workstation
  - 3270 BSC
  - 3270 SNA
  - '8100' SNA(PUT2)
  - 370X replacements
- DEC VT100
- Others
  - * Plug-compatible or added value?
  - * Compare with protocol converters

Figure (1)5: 'Plug-compatibles'

potentially different in some respect, product solutions need careful evaluation in order to understand the limitations of use and constraints they might impose.

> *Dewis:*
> *Companies should not be tempted to exploit technology for technology's sake, but should use it only where it can be seen to support their business needs. Examples of such temptations are all too clearly seen in the area of office systems, where many companies are currently installing individual modules which do not form a consistent whole.*

The final choice of *de facto* or *de jure* standards has to be made in the light of requirements and possible solutions. The important thing is that the decision process is broadly based (ie subject to a strategic plan) rather than narrowly based.

The need for a strategy

The process of integration addresses many facets of the work of an organisation, namely:

- Telecommunications (voice, data, text, image)
- Information processing (EDP, word processing, messaging)
- Corporate aspects (information flow, structure, O and M)
- Business.

(See Figure (1)6.)

It would be foolish to suggest that there is either a quick or a single solution to this problem. There is not. It depends on the nature of the organisation, its business and the particular requirements for information flow. It also depends on available funds for investment in change and a corporate willingness to develop a strategy to achieve some, if not all, of the possible goals.

> *Dewis:*
> *Different business needs require different strategies and therefore it should not be assumed that*

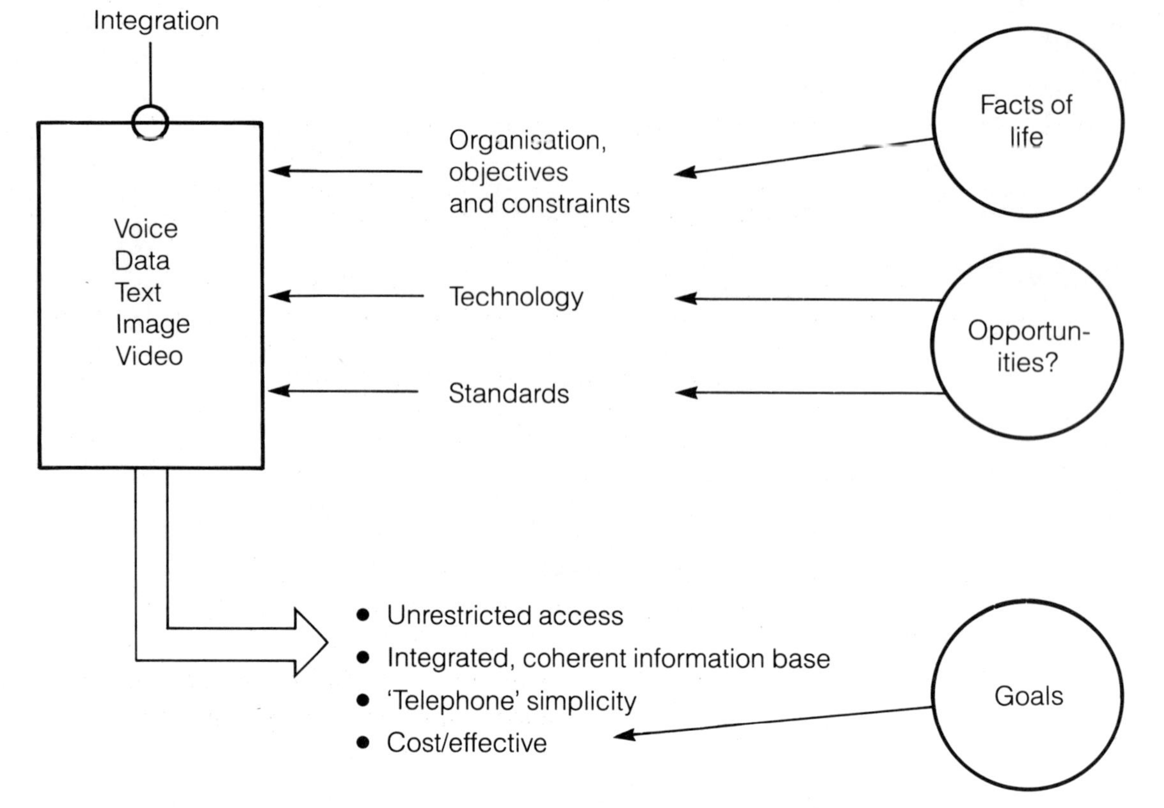

Figure (1)6: The process of integration

the strategy detailed here is directly relevant to any other company, but experience gained at BL Systems Ltd, renamed Istel Ltd in January 1984, shows that all companies should be thinking of developing their own equivalent strategies.

Eventually it must be realised that without a strategy, any partial integration that occurs will do so almost by accident (eg the local realisation that a leased telephone line could be shared for voice and data). This will yield suboptima which may be incompatible with a more global optimisation. The latter can only be obtained by strategic planning.

In the following sections of this Report, the technology and standards scenes are reviewed in order to educe the likely possibilities for integration. Finally, some thoughts are set forth on how a strategy might be developed within an organisation.

2: Voice and data

The world's biggest and most widespread network is the telephone network. Despite the growth in data communications, electronic message systems and television, the telephone continues to dominate the scene. This presents a dilemma, since the traffic patterns of these 'non-voice' services differ from those of the telephone. Yet there are strong economic reasons for attempting to combine these networks, at least at the basic transmission level. The trend in voice systems is away from analogue and towards digital transmission and this augurs well for integration in the future.

Voice and data

Combining voice and data

The telephone is the world's most common telecommunication device. The ubiquity of the telephone network has also meant that for many years it has also provided the cheapest, or indeed in some cases the only, way of providing data transmission capability for computers and their terminals.

> *Ettinger (1)*:*
> *The modern telecommunication network allows its users to connect some 500 million users to each other: the parties may be next door to one another or many thousands of miles apart. This remarkable world-wide system is the end-product of over 1000 billion dollars (at 1983 prices) of accumulated investment in the last 20 years by the telecommunications carriers all over the world and the willingness of these carriers to cooperate with each other in matters of tariffs and standardisation.*

This is not to ignore the importance of telegraphy, message-switching and telex networks but they cannot compete with the greater data transmission rates available and the sheer penetration of the telephone network.

Figure (2)1 shows the results of a survey carried out in the US together with a short-term prediction for the numbers of installed computer and communication devices *(006)*. This certainly bears out the statement about the predominance of the telephone, but it also reveals two other interesting trends:

1 The installation of domestic units exceeds that of business telephones, although the latter certainly offer much the greater traffic load. This has come about because economies of scale have brought the cost of the service within the reach of a wider section of the population and this in turn has reinforced the economics of business use of the telephone. This phenomenon may apply in other areas in due course.

2 The projected growth in desktop computers far outstrips growth in other areas. The ratio of desktops or personal computers to telephones will increase to such a point that it will be almost as common to have some such device on the desk, as it is to have a telephone instrument.

At this point it is interesting to note that there are two counter-operating pressures which are complicating the network scene. Firstly, the increasing use of the computer has led to the requirement for specialised data transmission and computer networks. The patterns of traffic generated by computers and their terminals are various and many and are quite different from those typically found in the telephone network. The Public Switched Telephone Network (PSTN) has been engineered and optimised to handle

---

*'Ettinger (1)' refers to the first of Ettinger's papers in the Invited Papers section.

	1981	1986
Mainframes	61	82
Minis	600	1700
Desktop computers	1200	14 600
WP/electric typewriters	1000	5200
Terminals	6900	17 000
	9761	38 582
Business 'phones	55 000	70 000
Home 'phones	130 000	145 000
TVs	80 000	90 000
Desktops/business 'phone	1:46	1:5

Figure (2)1: Numbers installed (thousands) USA

voice calls, particularly those of the business community. It may not be considered appropriate either to distort the telephone network operation to accommodate data requirements (which in any case currently represent only a small percentage of the total), or to expect the growth of computer network technology to be hampered by the (unnecessary) constrictions of the telephone system. This has led to the emergence of new and separate technologies for data networks such as packet-switching and local area networks.

In the second place, despite the appearance of separate networks for data requirements (eg the public data networks like PSS), there will be an increasing juxtaposition of a data terminal (for human use) and the telephone instrument. Indeed there are already devices on the market which combine the two (eg the STC executive telephone which, in addition to the handset, also has a full keyboard and a small integral screen).

This leads to pressures to combine the telecommunication provisions for the two types, voice and data (see Figure (2)2).

Levels of integration

In the public network domain functionally separate networks have emerged (telegraph, voice, data). In fact, although these networks are functionally independent, ie they have separate switching functions and separate terminals, they do share the fundamental transmission bearers (see Figure (2)3).

Transmission plant, be it cable, microwave, optical fibre or satellite, is installed to provide 'bandwidth' on a route between terminating or switching points. This capacity is divided into channels by means of multiplexing techniques which allow a multitude of voice or data streams to be carried simultaneously.

> *Ettinger (1):*
> *The past 70 years have seen remarkable developments in transmission technology that have resulted in a manifold reduction in costs per voice channel for long distance calls. The primary factor behind this success in the lowering of costs has been the development of systems*

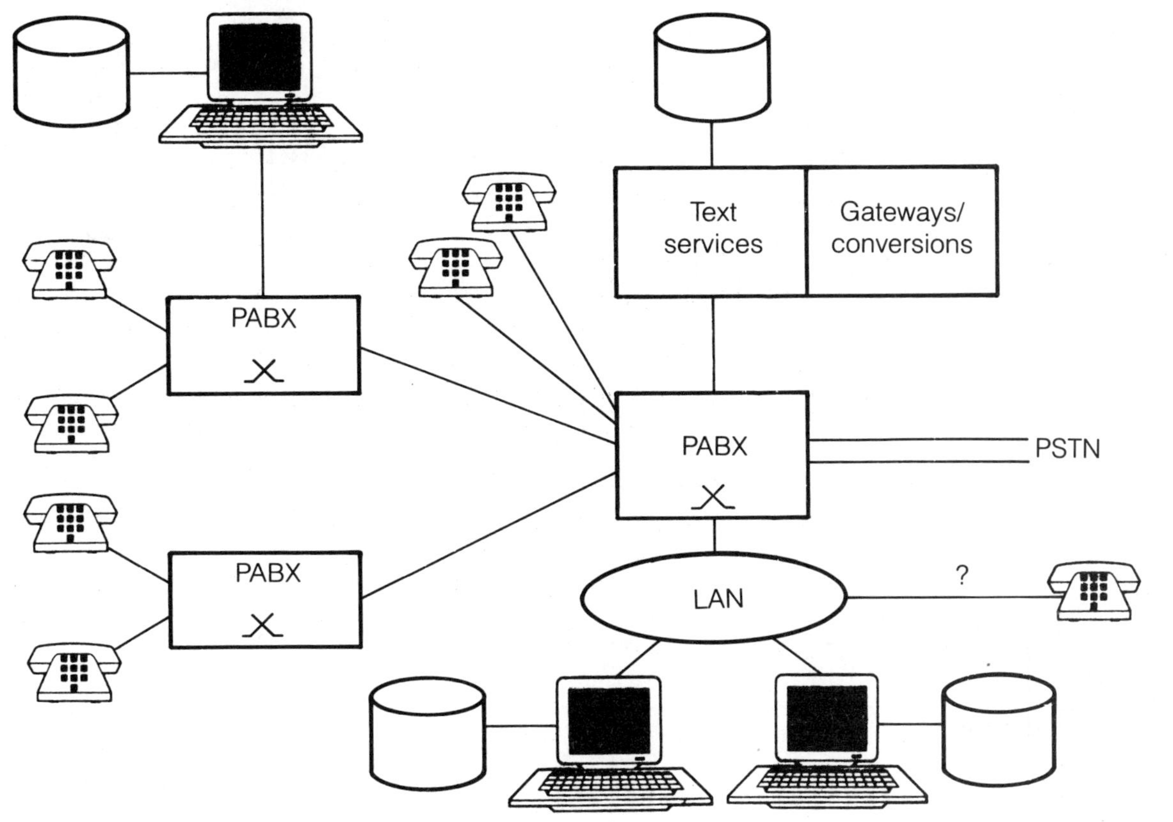

Figure (2)2: Voice-oriented evolution

of higher and higher capacity using terrestrial cable — coaxial or optical fibre, radio and satellite systems.

Shields:
Optical communications will certainly be the technique used for all new systems; it will also be used to extend the capabilities of our current installed transmission systems....

...for example, bandwidths of 10 000 GHz become feasible. This promises to enlarge the information handling capability of the optical signal.

However, as indicated above, because of the different traffic profiles these different streams are separated and switched independently.

In passing it may be noted that because of the predominance of voice, many of these transmission systems are designed to carry analogues of the voice signal and therefore have to be adapted for data transmission (eg by the use of modems). In the last few years, the introduction of digital techniques has accelerated the movement from the 'analogue' to the digital mode of transmission which is, incidentally, more suited to supporting data traffic (see Figure (2)4).

More recently the public network authorities have revealed that they plan to introduce an Integrated Services Digital Network (ISDN), bringing to the subscriber a single network termination over which a variety of services (voice, data, message, facsimile, image) can be provided *(007)*. Just how these services will be integrated at the functional level has not yet been determined.

Ettinger (2):*
In addition to the technical and economic arguments for ISDN, thepresent convergence of computing and telecommunications is creating a market for telecommunication services in

*'Ettinger (2)' refers to the second of Ettinger's papers in the Invited Papers section.

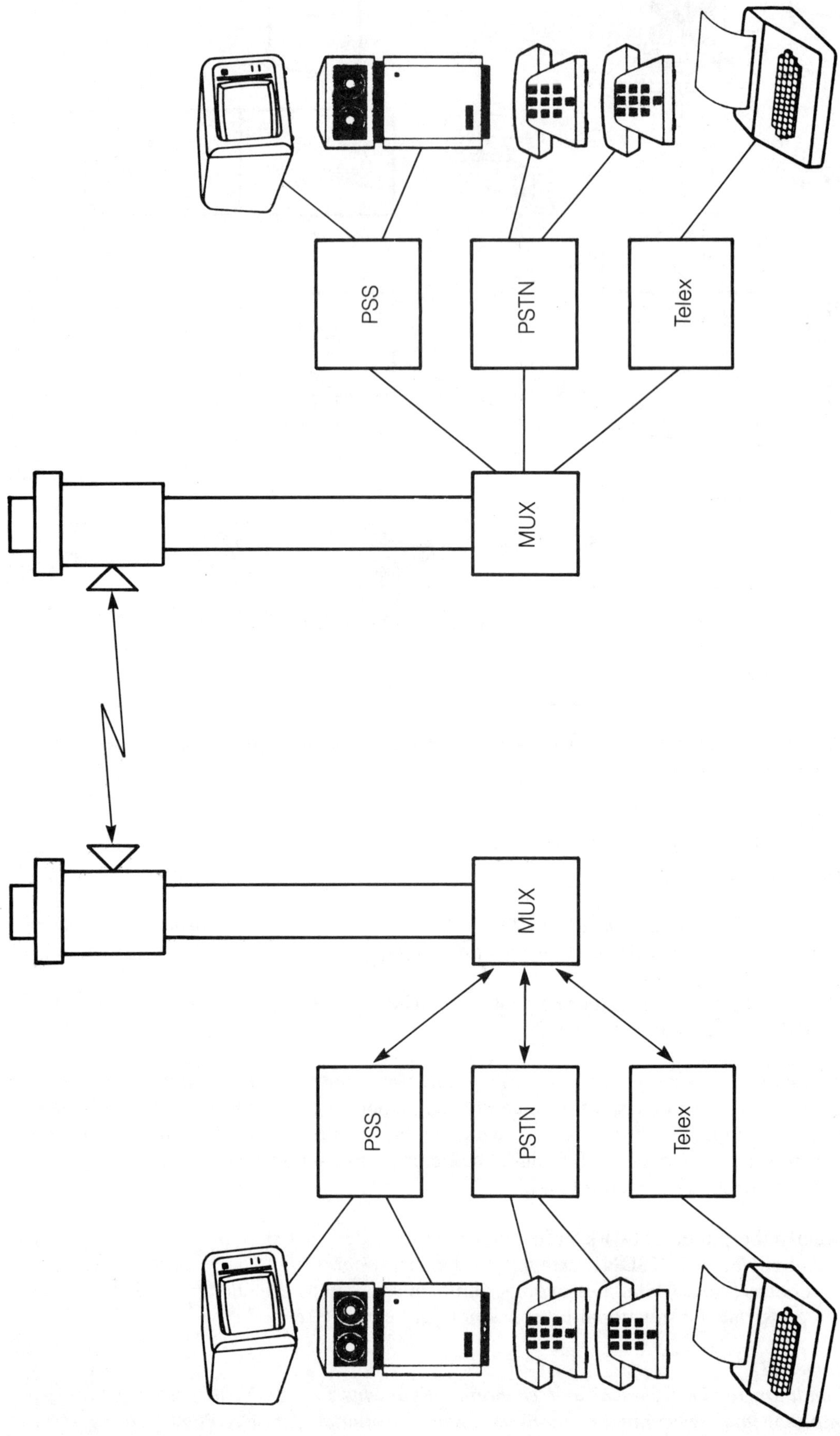

Figure (2)3: Common transmission bearers

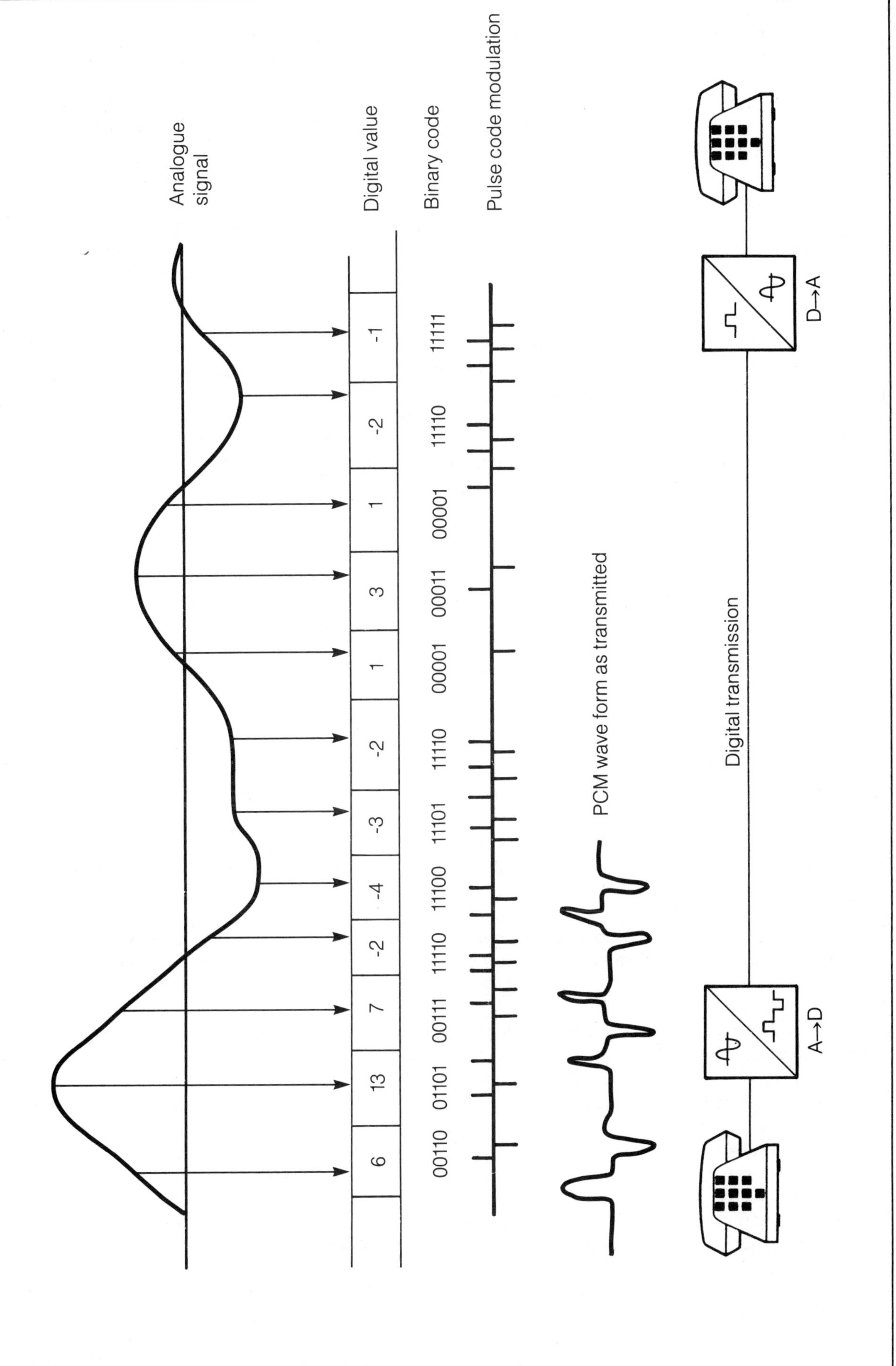

Figure (2)4: Digital transmission of analogue signals

general, and for integrated services in particular. The demand for integrated services arises from users' needs for telecommunication installation offering flexible access eg the Integrated Services Private Branch Exchange (ISPBX) and the Local Area Network (LAN).

Private networks

A similar picture has emerged in the private network arena. Large companies, having significant amounts of internal telephone traffic, have installed private telephone networks. In some cases this has simply been the use of leased tie-lines between buildings, allowing the local PABXs to be interconnected. More ambitious networks have included full tandem switching and unified network numbering plans.

Where there are also data network requirements it has also been possible to combine the voice and data requirements to some extent. As with the public network scene, this combination has frequently been just at the basic transmission level. In the British Steel network, for example *(008)*, the major sites around the country have been connected together with wideband circuits leased from BT. These circuits are divided into a number of voice channels for interconnecting the various local telephone systems plus the high-speed (48 Kbit/sec) data links for interconnecting the packet switches at the various sites (see Figure (2)5).

This network has been in operation for a number of years and was constructed using analogue transmission plant. Today the choice would probably be to use wideband digital plant such as BT Megastream 2. Products are starting to appear which facilitate the use of these new digital offerings (eg the Timeplex and Megamux multiplexors *(009,026)*) but it is still likely that the voice and data channels so derived would be connected to separate voice and data systems on the user sites.

The PABX and data

It has always been possible, by the use of simple modems, to put data calls through a PABX. In the past this facility was limited by a number of technical constraints inherent in the typical analogue PABX. These were as follows:

1 Most PABXs were 'two-wire', making both-way simultaneous data transmission at speeds above 300 bit/sec difficult (if not impossible).

2 Few, if any, PABXs had 'non-blocking' switch architecture. This meant that a few long hold-time data calls could spoil the Grade of Service for normal voice users (ie number of calls blocked by lack of switch capacity or lines).

3 The electromechanical environment (frequently found in older PABXs) is 'noisy', giving rise to high error rates for data calls.

4 If acoustic couplers are used, the telephone is blocked for the duration of the data call and cannot be used for voice.

Clearly the following are advantages in considering the use of the PABX:

1 The block-wiring of the building is already there; provided it can be shared with data, there is no need to provide separate and costly wiring for a data network.

2 Most offices have at least one telephone or telephone outlet in them. This makes the use of an acoustic coupler or integrated modem/telephone simple to provide and use.

This has led a number of PABX manufacturers to look seriously at the possibility of designing their new digital switches so as to reduce the above-mentioned problems and make the handling of data more attractive.

a)

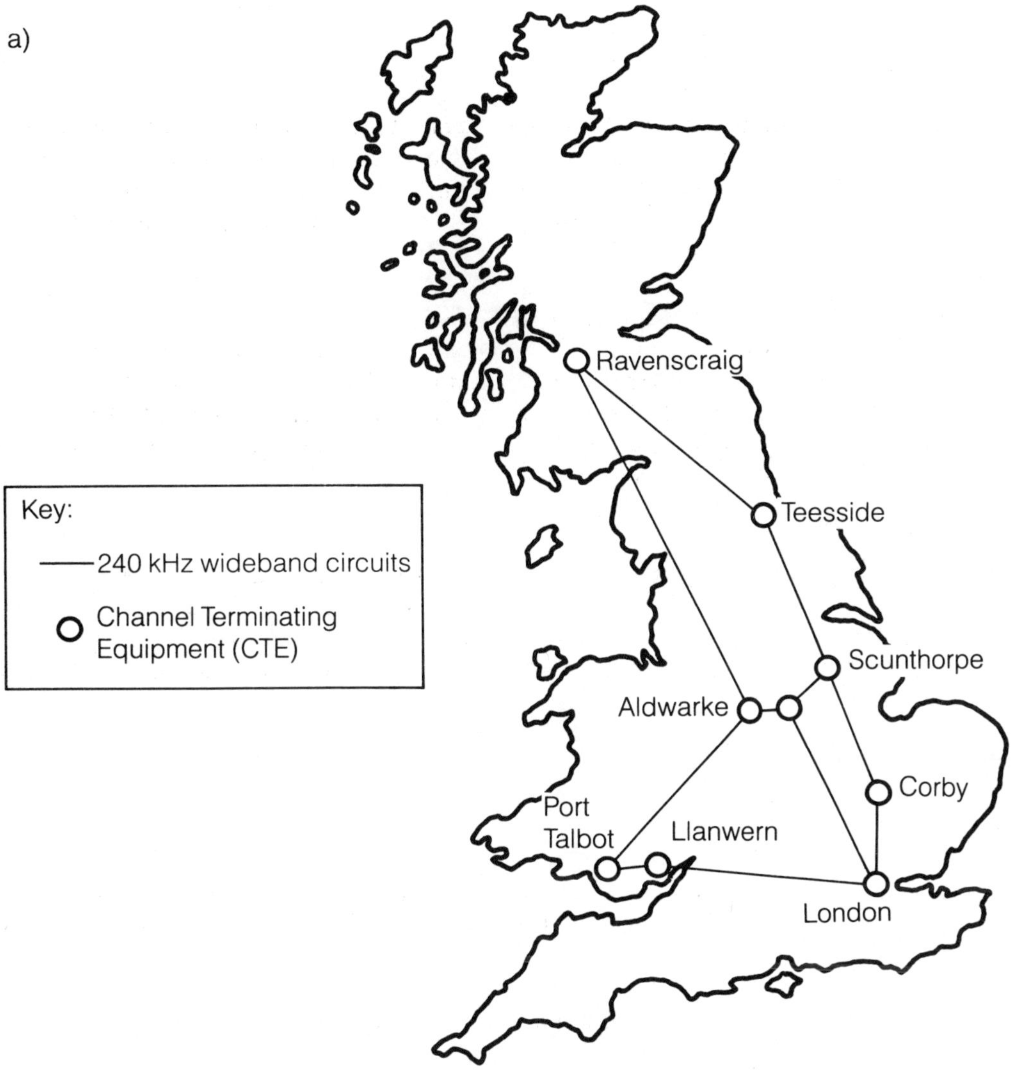

b)

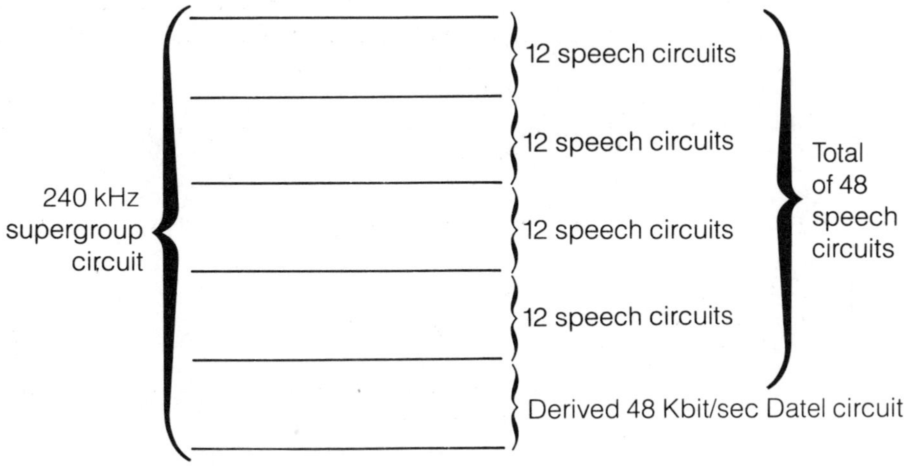

Figure (2)5a: Supergroup topography
Figure (2)5b: Use of post office supergroup circuits
Figure (2)5: British Steel Network

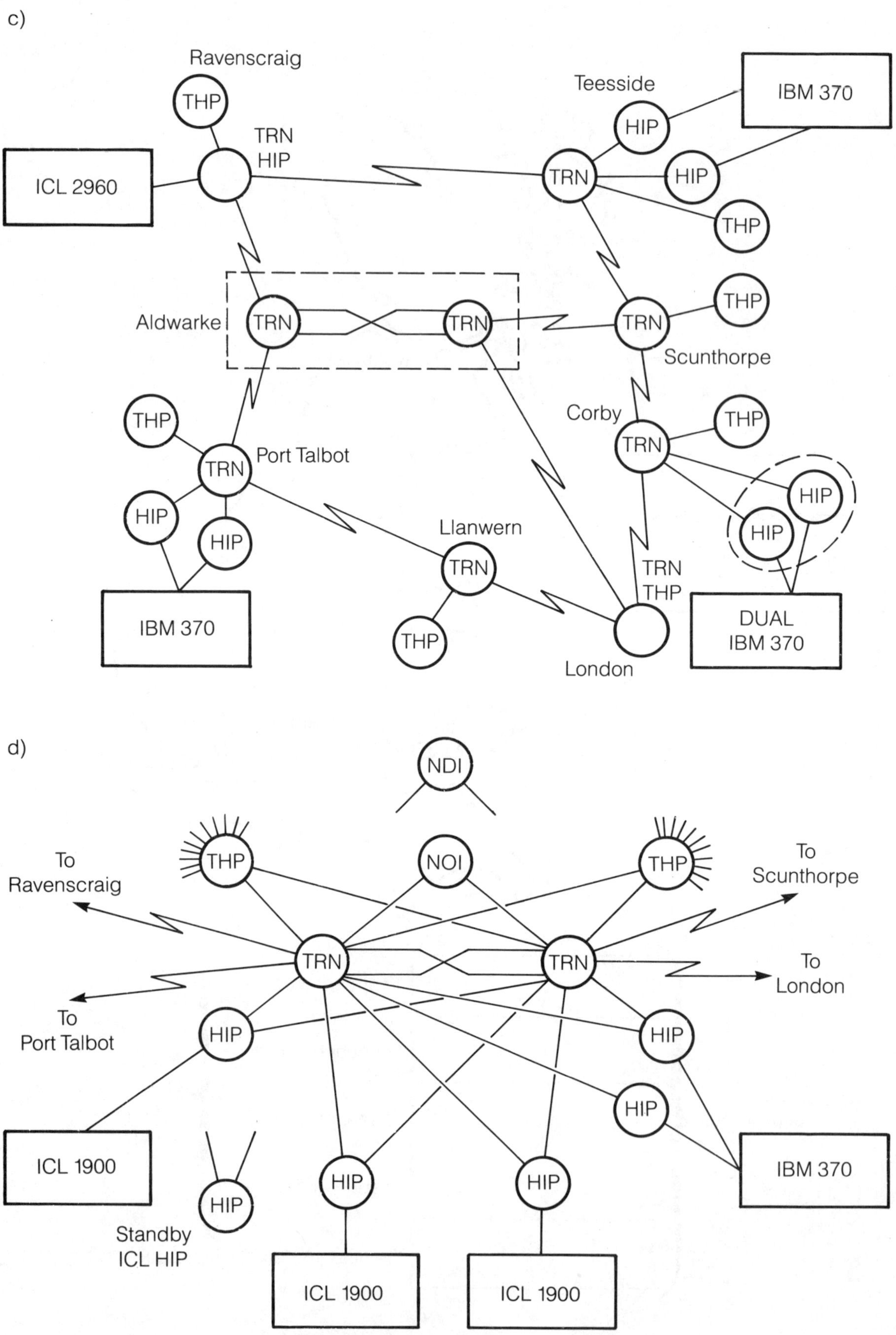

Figure (2)5c: Network topography (September 1978)
Figure (2)5d: Aldwarke topography (September 1978)
Figure (2)5: British Steel Network (cont)

The digital PABX

The architecture of digital switches (which use time-domain rather than space-divided switching mechanisms) makes it less costly and therefore much easier to provide a high Grade of Service to voice users in the presence of long hold-time data calls. Also it is possible to use the telephone extension wire-pairs to carry voice and data signals simultaneously thus avoiding the blocking of the telephone extension during a data call. (The telephone and the data terminal each use a separate 'port' and extension number on the PABX, of course.) The fact that digital telephony, as standard practice, uses a 64 Kbit/sec data rate to carry digitised voice, means that it is quite feasible to carry data at rates up to this value through the PABX.

On the surface, therefore, the use of a digital PABX as the basis of the integration of voice and data communications looks quite promising. Unfortunately, as with many such things, there are a number of countervailing considerations, as follows:

1 High-speed data transmission in the digital mode cannot at present be extended into the public telephone network (ISDN and universal access to the System-X or the digital network are some years away yet).

2 Modems will still be required for off-site data transmission (unless digital services such as Kilostream are available — and then this is only on a point-to-point basis).

3 Interworking between different manufacturers' PABX products will present some problems since the requisite standards are not yet in place (and may be impossible to 'retrofit' to existing installations). This may mean that within an organisation all PABXs would have to be supplied by the same manufacturer.

4 Circuit switching is not necessarily the best approach for multiuser computer systems. The provision of significant numbers of ports/extensions in order to support dial-in terminals is both costly and inappropriate. Leased-line and multiplexor, or packet-switched approaches are increasingly used, particularly where intelligent workstations are considered. Few PABXs, if any, currently offer X.25-packet mode interfaces.

5 The digital voice rate of 64 Kbit/sec is curiously inappropriate for data. It is unnecessarily high for many simpler terminals but not high enough for intelligent devices operating in the close-coupled LAN mode (where megabit data rates are indicated).

Despite these considerations PABX manufacturers are pushing ahead with products which may be attractive to some organisations (ie map onto their particular requirements). For example, Plessey are basing an office automation product, IBIS, on their digital PABX system IDX *(010)* (see Figure (2)6).

Other non-voice facilities

The predominance of voice has led to the situation where everything else is referred to as 'non-voice' requirements (usually under the general label of 'data').

> *Burren:*
> *If challenged on this state of affairs, communications engineers will remind their computing colleagues that, in spite of all the fuss and excitement generated by computer networking, it is the telephone that is still by far the dominant source of telecommunications traffic.... There is an easy answer: in time the telephone will become a digital device and will itself be computerised.*

While the emergence of public data networks like PSS has focused attention on computer-oriented applications, it must not be forgotten that other types of traffic exist, such as:

- Telex and message-oriented systems
- Facsimile
- Video.

Indeed, whereas voice is comparatively profligate with bandwidth when compared with many 'data' or message applications, full-colour moving-picture video requires about three orders of magnitude more bandwidth than a telephone call. Fortunately there appears to be little requirement at present for two-way simultaneous video services (most are presently broadcast). Video-conferencing has not proved popular

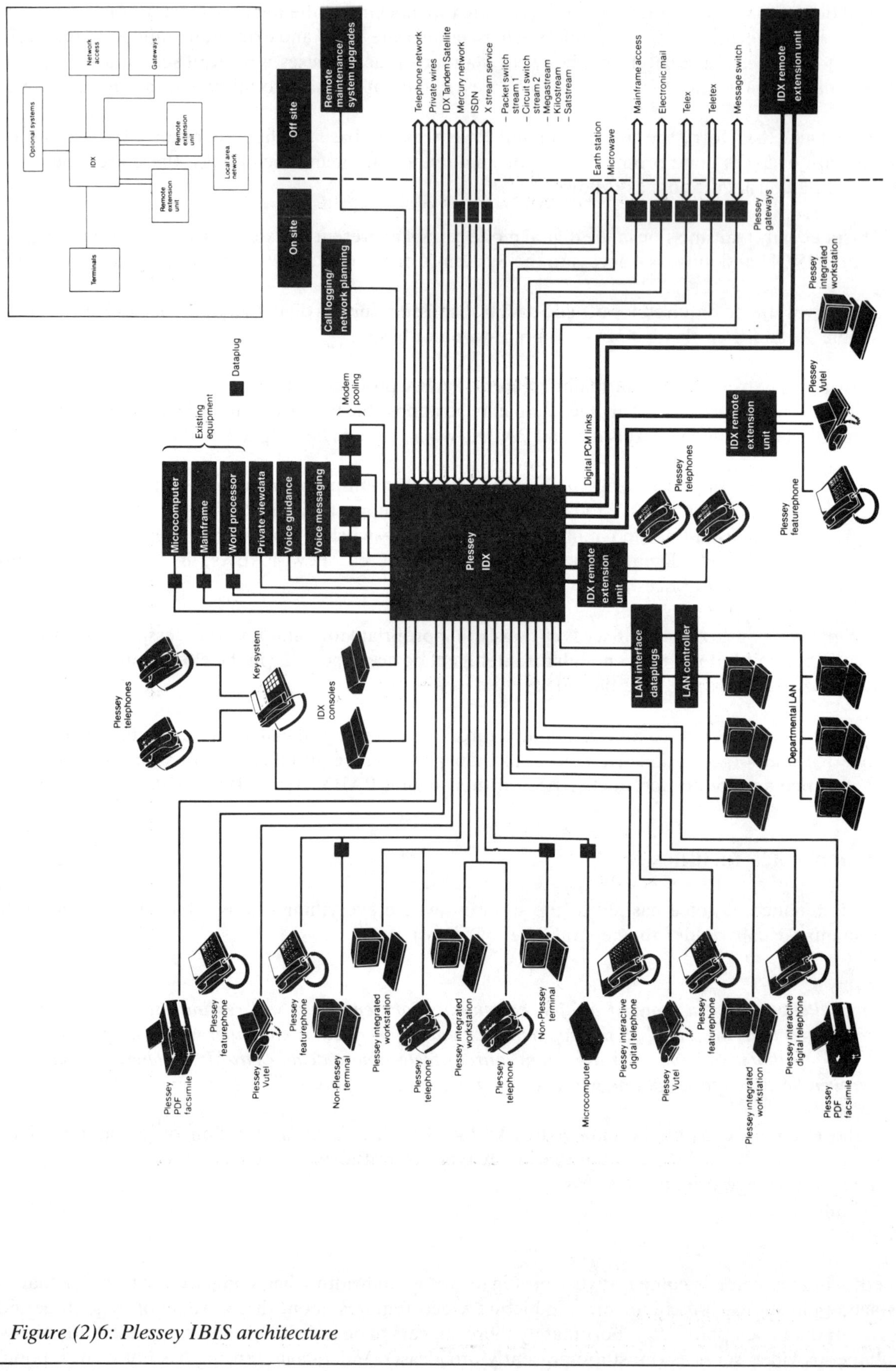

Figure (2)6: Plessey IBIS architecture

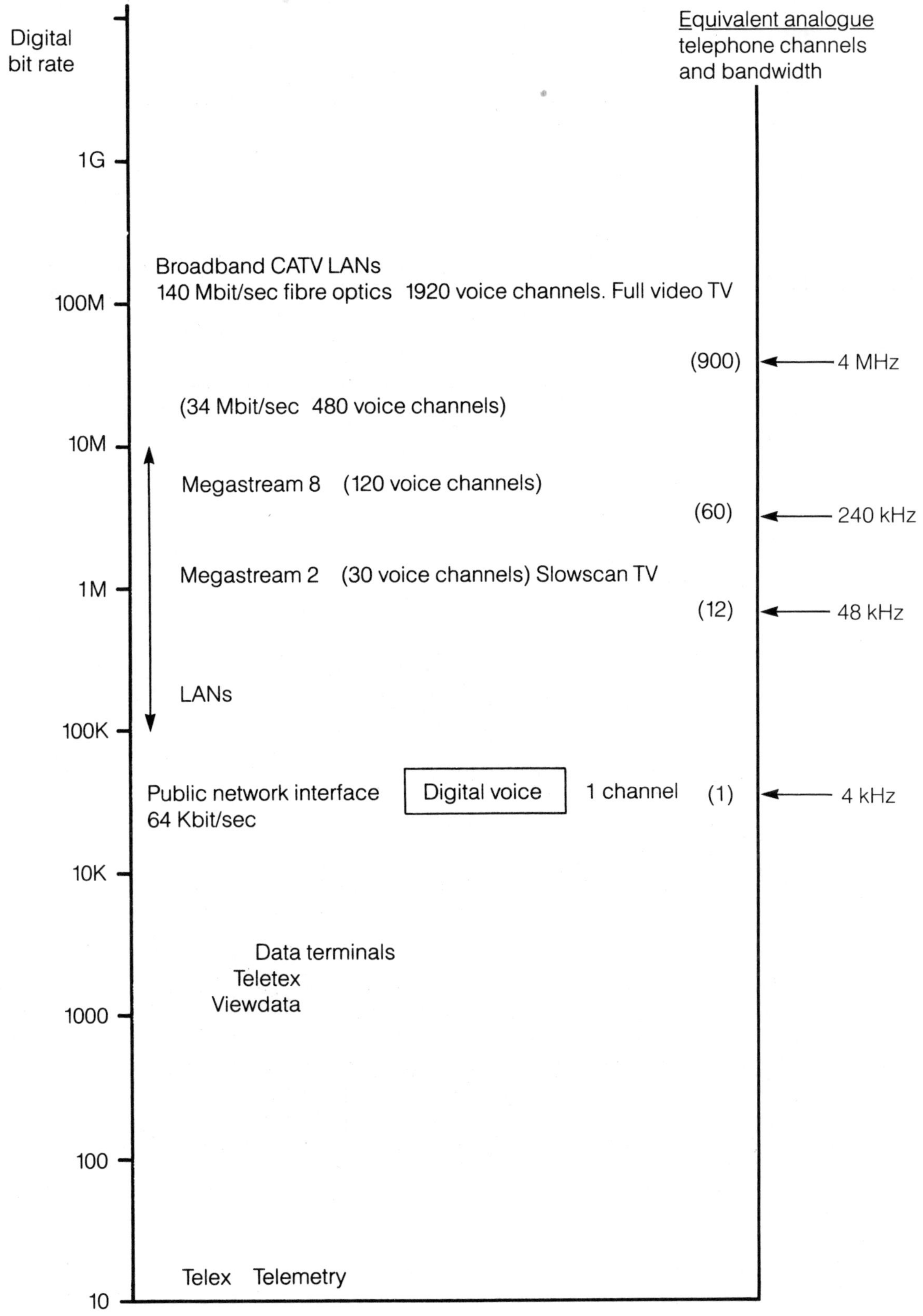

Figure (2)7: Digital data rates

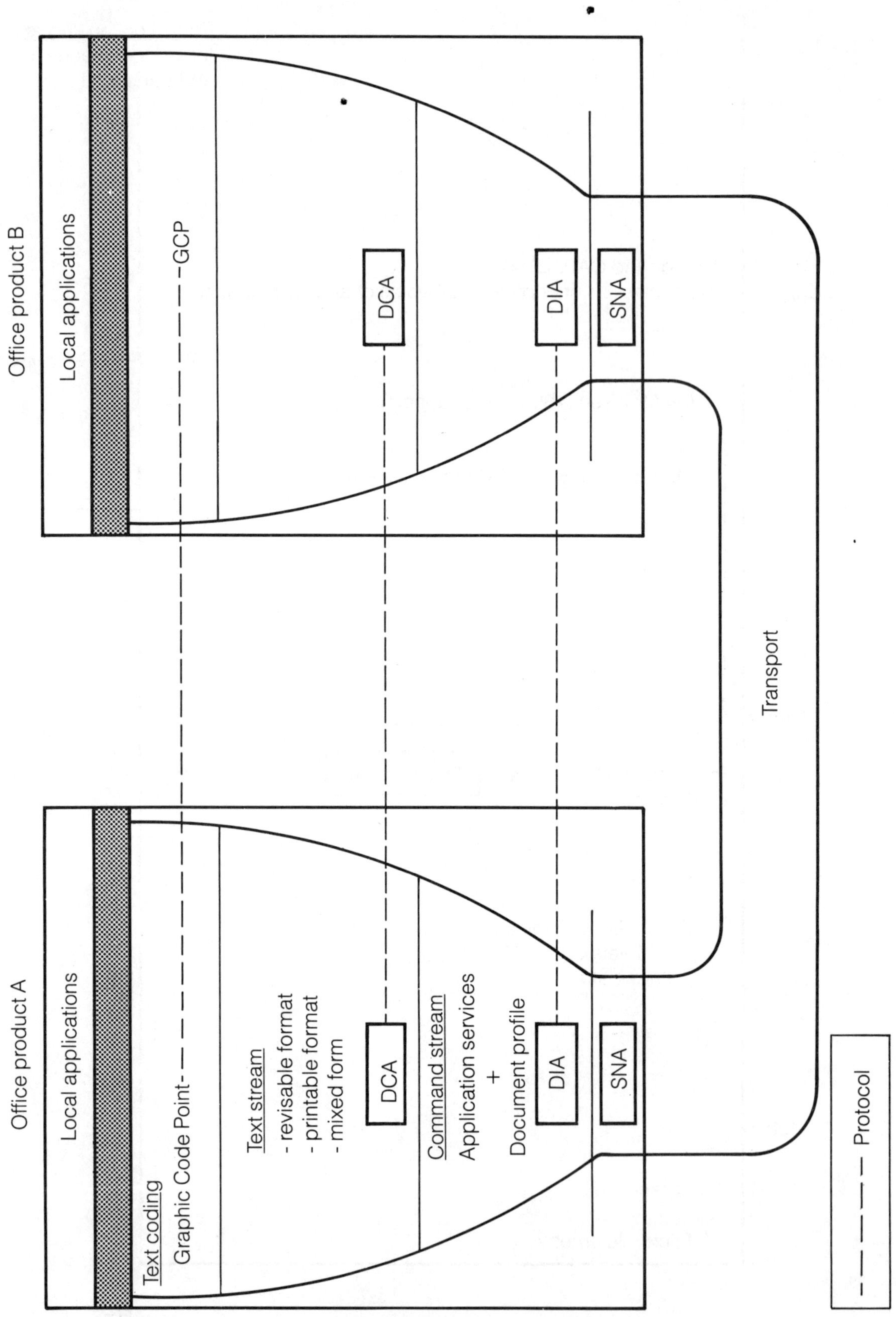

Figure (2)8: IBM communications architecture for office products

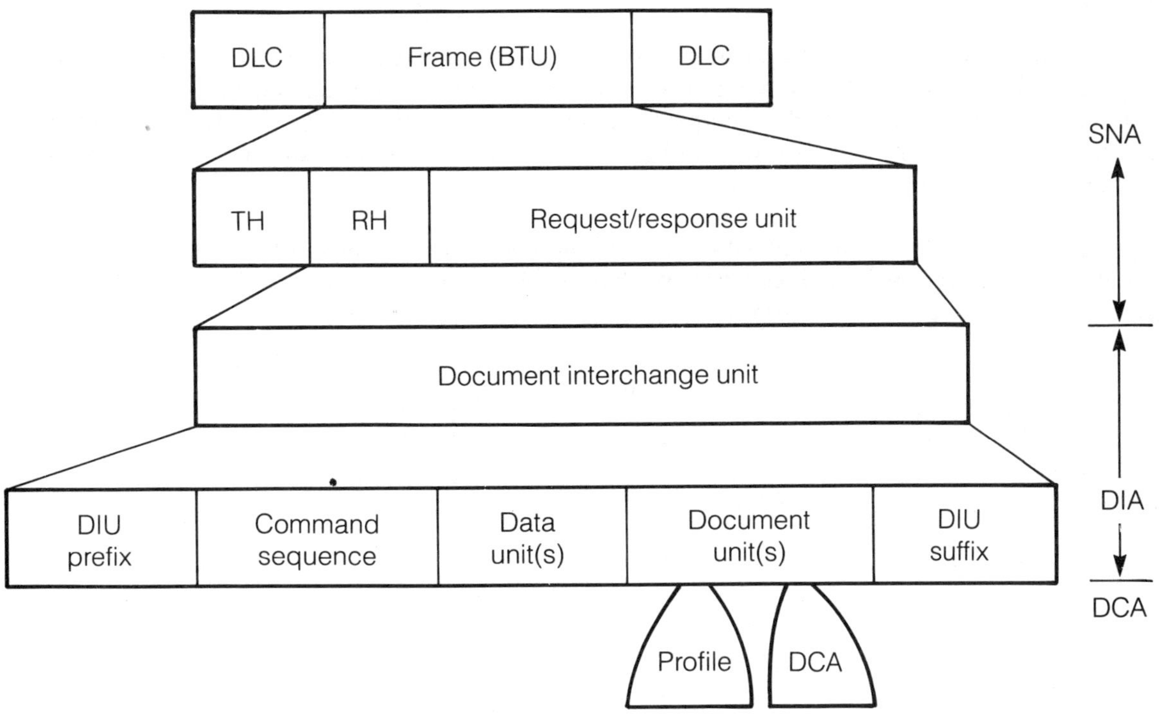

Figure (2)9: Document information flow

and neither has 'view phone' which uses slow-scan TV (equals about 30 voice channels). Still-pictures, on the other hand, can be encoded for transmission (videotex) using more modest bandwidths in line with those used for data transmission (see Figure (2)7).

> *Burren:*
> [Burren's] *Figure 1 lists some sources of traffic and gives equivalent volumes in terms of digitised voice at 64 Kbit/sec. For example, one A4 page of text is equivalent to half a second of voice and 'conferencing' quality colour TV is equivalent to about 30 voice channels.*

In fact, with the present exception of facsimile, most non-voice services such as viewdata, message systems, Teletex, electronic mail and other 'office automation' facilities can be conveniently placed together with 'data'. Perhaps the fact that they are all increasingly based on computer systems has something to do with this.

> *Burren:*
> *One would in some circumstances operate the system as a real-time 'conversation' copier to copier. This mode would be useful in, for example, a teleconference situation, but in general it would be a great nuisance to have to synchronise the use of two copiers at different locations. Thus control computers with storage are required to organise the flow of data.*

> *Burrows:*
> *... the main development of Prestel has been to create a new market not covered by on-line systems. This is mainly transitory information in the following areas:*
> - *Hotel booking*
> - *Travel — tourism*
> - *Banking — electronic funds transfer*
> - *Teleshopping*
> - *House purchase — estate agents*
> - *Advertising*
> - *Education*
> - *Community information.*

Even facsimile is changing. Present analogue systems, such as group 1 and group 2 machines used on the telephone network, are being superseded by the new group 3, and ultimately group 4, digital machines — the latter promising to offer integration with text facilities like Teletex (see 'Text and information systems' in *(019)*).

This presages the widespread introduction of systems in which text and graphical (picture) information can be combined. The main question here is whether the standard can be agreed before it is overtaken by events and products, like IBM'S Document Content Architecture *(011)*. The IBM communications architecture for office products is shown in Figure (2)8. Figure (2)9 illustrates an example of document information flow.

These considerations serve to highlight the voice/non-voice dichotomy and the struggle involved in attempting an integration. The question is at what 'level' is this appropriate and is a particular solution universal or only specific to the instance?

3: System architecture

The juxtaposition of communications and computer technologies in supporting the user and his 'applications', has given rise to the need for a coherent framework or model of the total system. This framework has been called the system 'architecture'. Manufacturers of computer systems have produced their own individual system architectures, which has led to incompatibilities between them. This is restrictive and can be costly for the user. International standards are now being formed which offer a potential solution to this problem.

System architecture

Network architecture

The foregoing sections can be paraphased as 'there are applications which require communications'. This statement is encapsulated in Figure (3)1.

Returning to the issue of network integration it can again be seen that this is not a single issue, even at the technical level, but involves several different technical issues, as follows:

- Transmission (analogue, digital, cable, radio)
- Switching (packet, circuit, message)
- Application types (EDP, WP, voice, image)
- Application support (hardware, software, operating system, filehandling, DBMS)
- Interworking (protocols, formats, layering)
- Management (fault logging, statistics, growth, cost allocation, security).

As mentioned above, integration presupposes some standardisation and the critical area here is the fundamental system architecture.

Although conceived in the arena of data processing systems, the Open System Interconnection (OSI) architecture as promulgated by the International Standards Organisation (ISO) is sufficiently general in concept and scope that it can encompass most, if not all, applications that we can currently hope or expect to consider *(012)*.

> *Tucker:*
> *OSI is a programme of development of specific protocols which are designed to have very wide applicability in data communications: it is not, however, the solution to all known problems. OSI is specific; it is not something that anyone can simply pick up and reinterpret or modify; nor can sections be extracted at will. If this does happen the result will not be OSI and should not be described as such. Furthermore, the protocols, and even the Reference Model, may not be appropriate in wider contexts. In particular, their applicability to voice and wideband communications deserves careful examination: it should not be a foregone conclusion.*

The OSI model describes an architecture with a total of seven principal layers. These can be grouped into the following three sets of functions:

1 Communication support.
2 Application support.
3 End-user applications.

This architecture is illustrated in Figures (3)2 and (3)3.

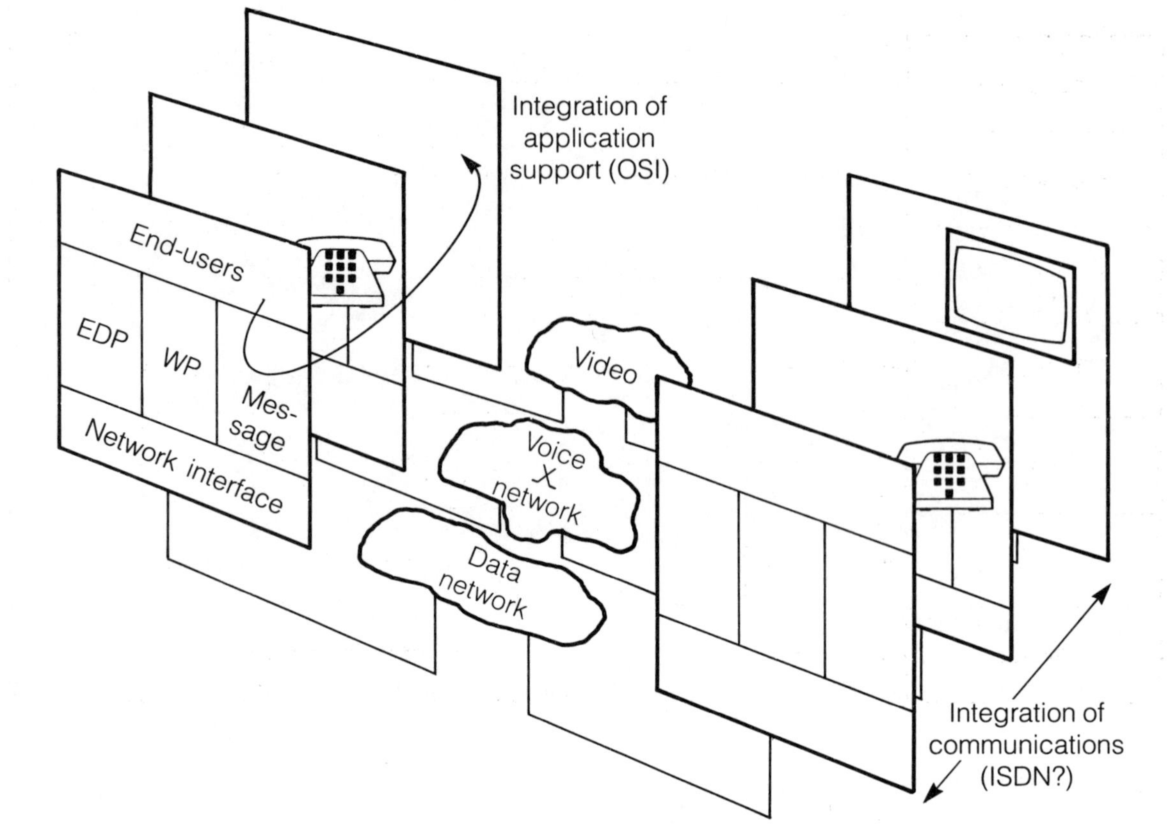

Figure (3)1: Dimensions of integration

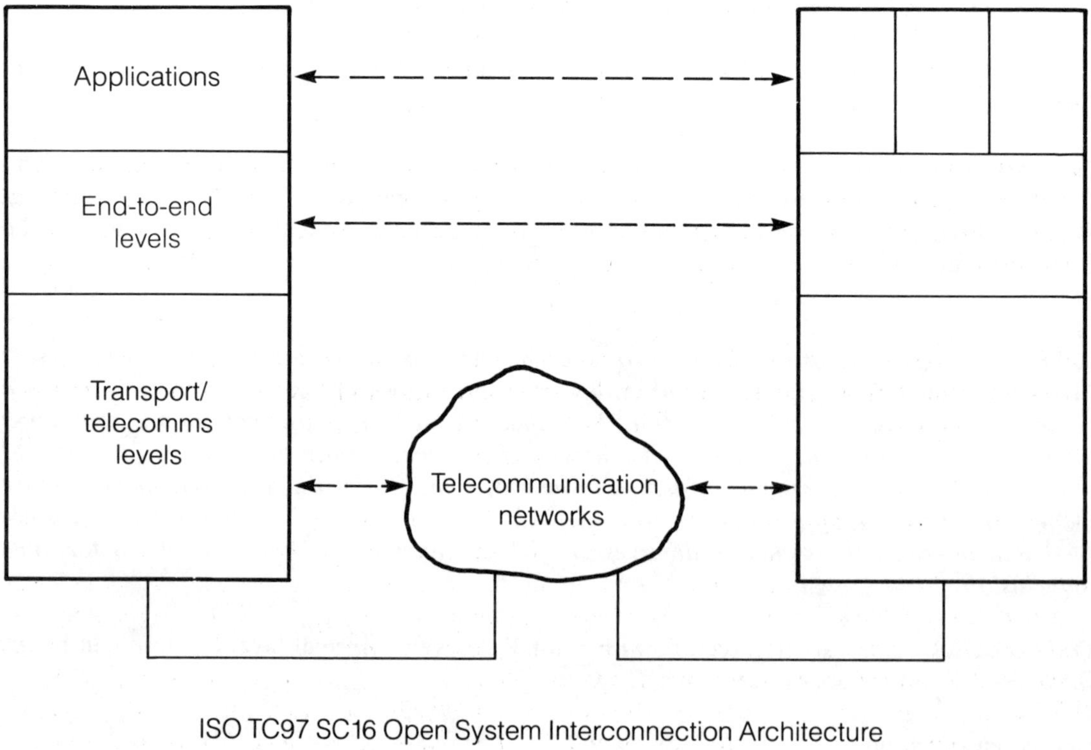

Figure (3)2: System architecture

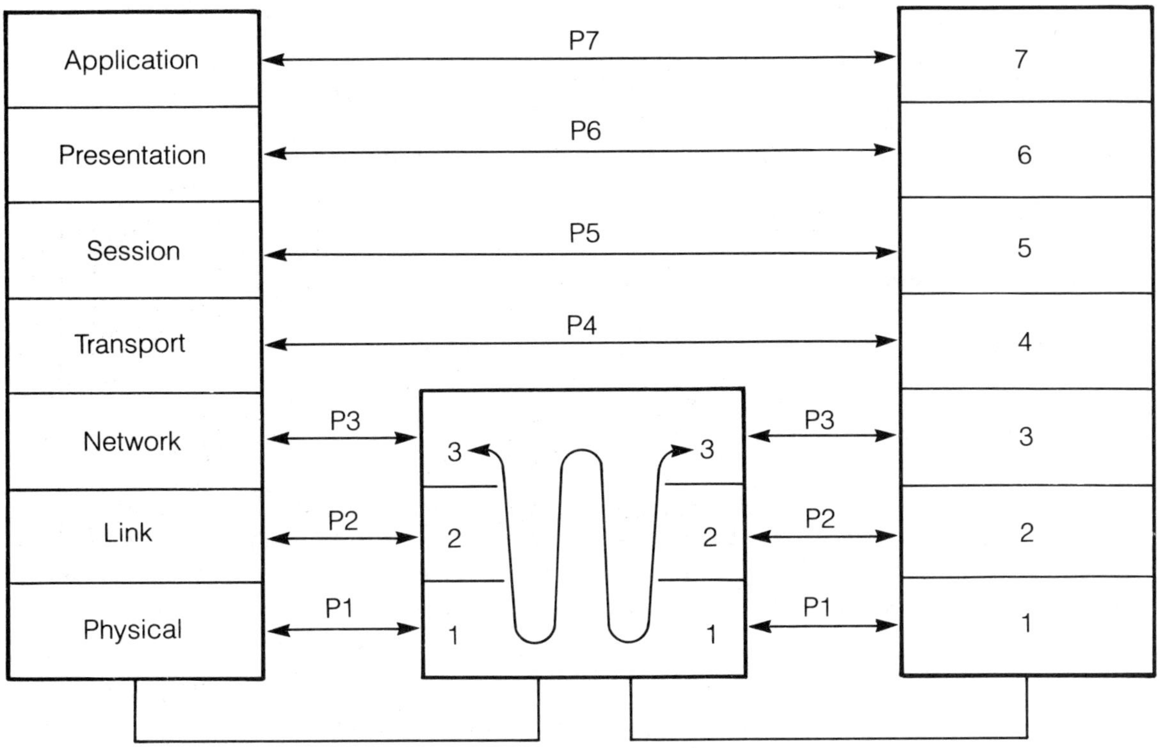

Figure (3)3: Open systems architecture

Tucker:
The agreements on these structural matters are documented in standards called the OSI Basic Reference Model and Layer Service Definitions. They are made standards in order to freeze the specifications, not for external consumption. They are of importance only to those developing OSI protocol standards: there is no sense in which systems can claim to comply with them.

The communication support layers (one to three) are involved (principally) in providing telecommunications handling. This includes the actual transmission and reception of bits across the physical link interface, the operation of any protocol across the link associated with grouping bits into transmission blocks or messages, and invoking the signalling procedures needed to operate a switched network where necessary.

In practice, various telecommunication network interfaces are available (eg X.25, X.21, V.25, IDA/ISDN etc) but they all provide different features and these will have to be 'interpreted' in some sense in relation to the ISO model (this point is discussed further in Section 4).

Tucker:
However, the OSI standardisation bodies will not be developing all the future standards for communications interfaces and networks, and so it has to be recognised that real networks will not necessarily be developed to use OSI protocols and may not even correspond in their allocation of functions to the Reference Model. Today, the most commonly-referenced network procedures are those of X.21 and X.25 in the public domain, with ISDN on the horizon and the IEEE 802 standards for local area networks. None of these corresponds exactly to OSI, although X.25 comes close. It is not even desirable that OSI protocols should be applied right at the bottom of the architecture since that would put constraints on innovation in communications technology. What is needed instead, is a defined method by which a variety of low-level communications procedures can be accommodated in a world of interconnected networks across which OSI can be used and its benefits obtained.... Clearly, different subnetworks will in general provide different subnetwork services. In principle, for any subnetwork service (in practice for

all the significant ones) one or more OSI network protocols will be standardised capable of 'bridging the gap' between the subnetwork service in question and the OSI network service. (The extent to which these protocols should be general as opposed to subnetwork-specific is an important Network Layer issue.) In effect, the idiosyncratic characteristics of particular networks will not be visible above the Network Layer, over which standard OSI protocols may operate. The Network Layer also has the task of interconnecting subnetworks in order to provide communications end to end: that again will not be visible to higher layers.

The application support layers (four to six/seven) are involved in bringing the end-user applications into the networking environment. They will provide both the support for invoking the telecommunications layers and the support for interworking between the end-user system environments.

Unfortunately, although draft standards have been produced for the Transport and Session Layers, there is still discussion taking place over the distribution of functions between the Presentation and Application Support Layers (see Figure (3)4).

Tucker:
Unfortunately, this central block of well advanced standards is of little use on its own, because it requires a network protocol (as previously explained) to be mounted on top of real subnetworks; and even if that were present, the session service by itself is not intended to meet application process needs. Standards for the other layers are catching up, but one will have to wait until 1985 for a set of firm standards which will meet real application needs over a reasonable variety of communications technology.

So far, three principal types of application have been identified, namely:

1 File transfer.
2 Virtual terminal.
3 Remote job entry and control.

	CCITT	ISO	ECMA	
1	V and X Series	Connectors (25,37,15, way)	(E80/81)	ETHERNET
2	LAP/B/D	HDLC	(E.82)	ETHERNET
3	X.21, X.25, X.213	X.25 + (DIS 8208)		
4	X.214, 224 (S.70) (S.100)	DIS 8072/3	E.72	
5	S.62	DP 8326/7	E.75	
6	S.61		File E84/5	
7	S.60		V.Term (E.88)	

Other: X.121 Document, MHS (X.400). FAX T4.

Figure (3)4: Standards (OSI IS7498, X.200)

Until the functional requirements ('services') are agreed, specification of protocols at these levels cannot proceed. Thus, whereas the broad principles of the OSI architecture have been widely agreed, it has taken, and is continuing to take, a considerable time to formulate and agree even a minimum set of 'interworking' standards. It is like a chain — until every link has been forged it is not complete and cannot be used.

> *Tucker:*
> *An important point to note is that the layers sit above one another in the sense that each particular layer relies on the operation of those below. They are not alternatives: each is indispensible.*

For each level, the services to be supported and the peer-to-peer protocol to be used, must be agreed. If different applications give rise to different requirements (eg file transfer, virtual terminal, remote job entry) then more than one protocol may be needed at a particular level (see Figure (3)5).

For the computer equipment manufacturers, the dilemma is what to do in the meanwhile. Most of the manufacturers have, of course, produced their own 'system architecture' products; these differ from each other both in layering and in the structure of the protocols used (see Figures (3)6 and (3)7 *(013)*).

> *Tucker:*
> *... OSI is not a general-purpose, descriptive tool that will provide an understanding of other sets of protocols. Occasionally the OSI Reference Model is (mis-)used in this way, but this only leads to confusion. The Reference Model is neither general purpose nor descriptive: on the contrary it is a special-purpose and prescriptive tool which makes specific allocations of functions to layers, thus deliberately constraining (ie part-specifying) OSI protocols. Other network architectures may partition functions differently, in which case the Reference Model fails to describe them.*

Proprietary architectures: SNA

Perhaps the best known proprietary architecture is IBM's System Network Architecture (SNA) *(014)*. This has been around, in concept at least, for some considerable time. It actually predates the OSI model — although in its main hierarchical principles it is very similar (see Figures (3)8 and (3)9).

> *Gurugé (014):*
> *It is becoming the vogue to put forward multiple, essentially germane, but nevertheless somewhat divergent definitions of exactly what Systems Network Architecture (SNA) is. For instance, the following descriptions have been given:*
> - *'SNA is a direction-setting concept'*
> - *'SNA is a set of products'*
> - *'SNA is a specification'*
> - *'SNA is a plan for structuring a network'*
> - *'SNA defines the behaviour of networks of heterogeneous, loosely-coupled processors'.*
>
> *SNA per se, as introduced to the computing fraternity by the International Business Machines (IBM) Corporation in 1974, and as the word 'architecture' in its name unequivocally proclaims, was, and still is, fundamentally and intrinsically a design specification.*

SNA is strongly 'host-computer centred'. This is hardly surprising since it has evolved in an area where the preponderance of the system support and expertise has always been concentrated at the mainframe. The evolution from the application-dependent (pre-VTAM) to the application-transparent environment has taken place initially within this framework. Following this, the concept of distributed functionality has been invoked, with firstly the increased functionality in terminal controllers and other such devices (eg 4700 and 8100) and secondly, the advanced communications features which allow the construction of networks with multiple host computers.

> *Gurugé (014):*
> *In view of the data communications-related drawbacks inherent in 'pre-SNA' systems, the overall intentions of SNA are easy to surmise: the specification of the data communications-*

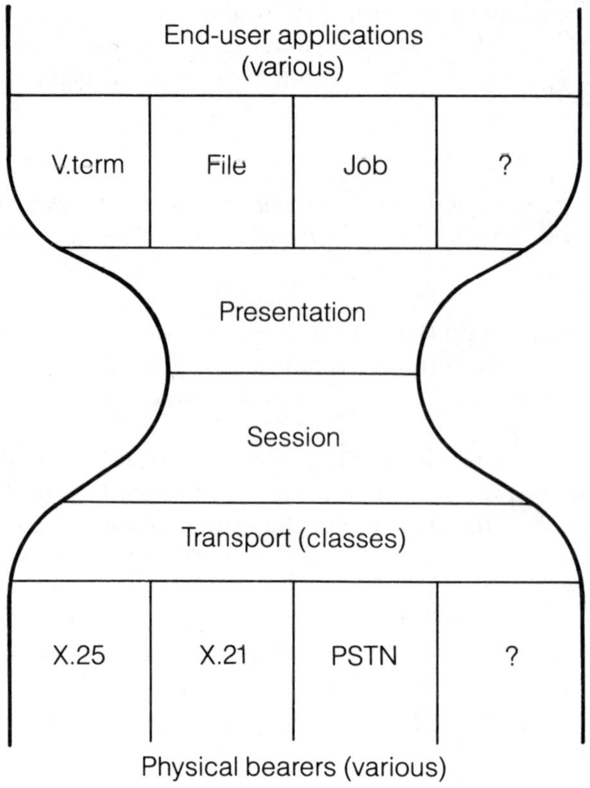

Figure (3)5: 'Hour-glass' architecture

IBM	Systems Network Architecture
Univac	Distributed Communications Architecture
Honeywell	Distributed Systems Architecture
NCR	Distributed Network Architecture
Siemens	Transdata
Burroughs	Burroughs Network Architecture
ICL	Information Processing Architecture
DEC	Digital Network Architecture
Hewlett-Packard	Distributed Systems Network
Modcomp	Maxnet
Prime	Primenet
Norskdata	Nordnet
Data General	Xodiac
Tandem	Expand
etc	

Figure (3)6: Manufacturers' products

Relation with the OSI reference model

OSI (International Standards Organisation)	OSI	SNA (International Business Machines)	OSI	DNA (Digital Equipment Corporation)	OSI	DCA (Univac)	OSI	BNA (Burroughs)	OSI	IPA (International Computers Limited)	Full XBM
7 Application	7	End user	7	End user	7	End user	7	End user	7	Application	
6 Presentation	6	Function management services	6	Network application	6	Communication system user Logical port	6	Host services	6	Dialogue control	Access levels
5 Session	5	Data flow control	5	Session Network services	5	Logical port multiplexor	5	Port level	5		
4 Transport	4	Transmission control	4		4	Data unit control	4	Network services	4	Telecommunications control	N/A
3 Network	3	Virtual and explicit route control	3	Transport	3	Routing control	3	Router level	3		Group level
2 Link	2	Transmission group control Data link control	2	Data link control	2	Trunk control Data link control	2	Station level	2		Link level
1 Physical	1	N/A	1	N/A	1	N/A	1	N/A	1		Physical level

Figure (3)7: Comparison of proprietary architectures with the ISO model

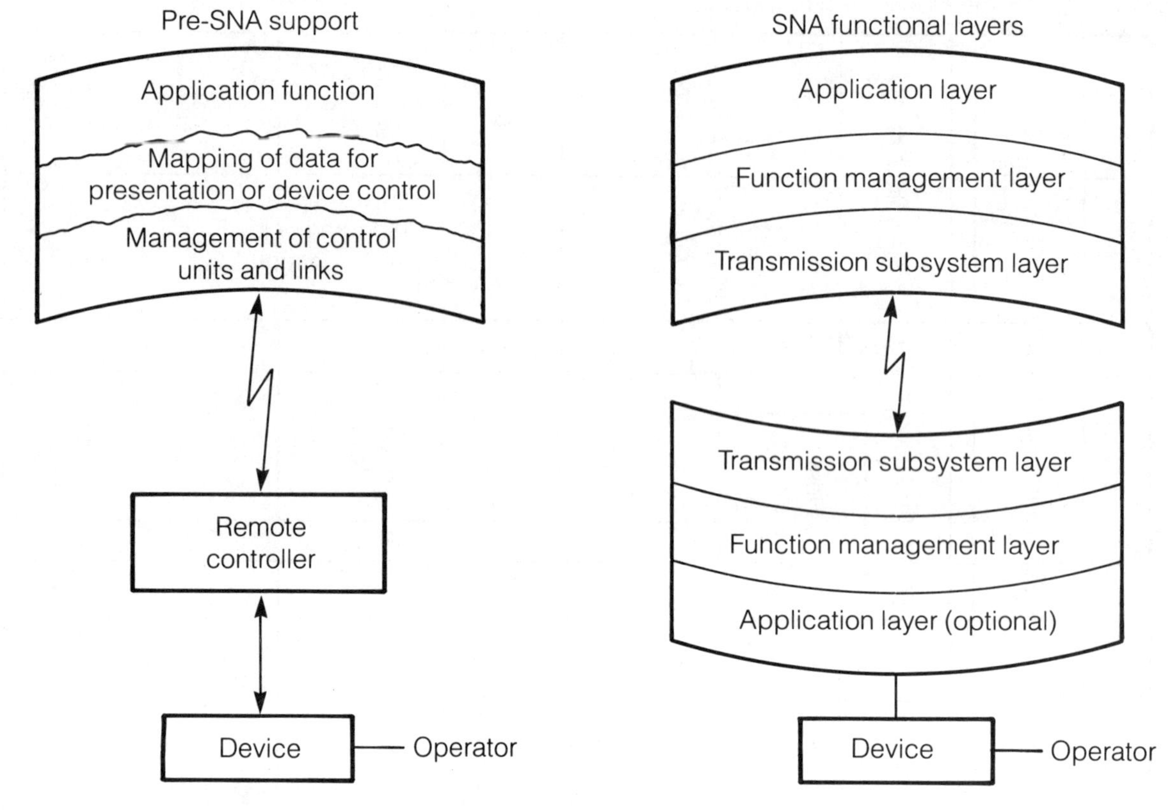

Figure (3)8: Functional layering - old versus new

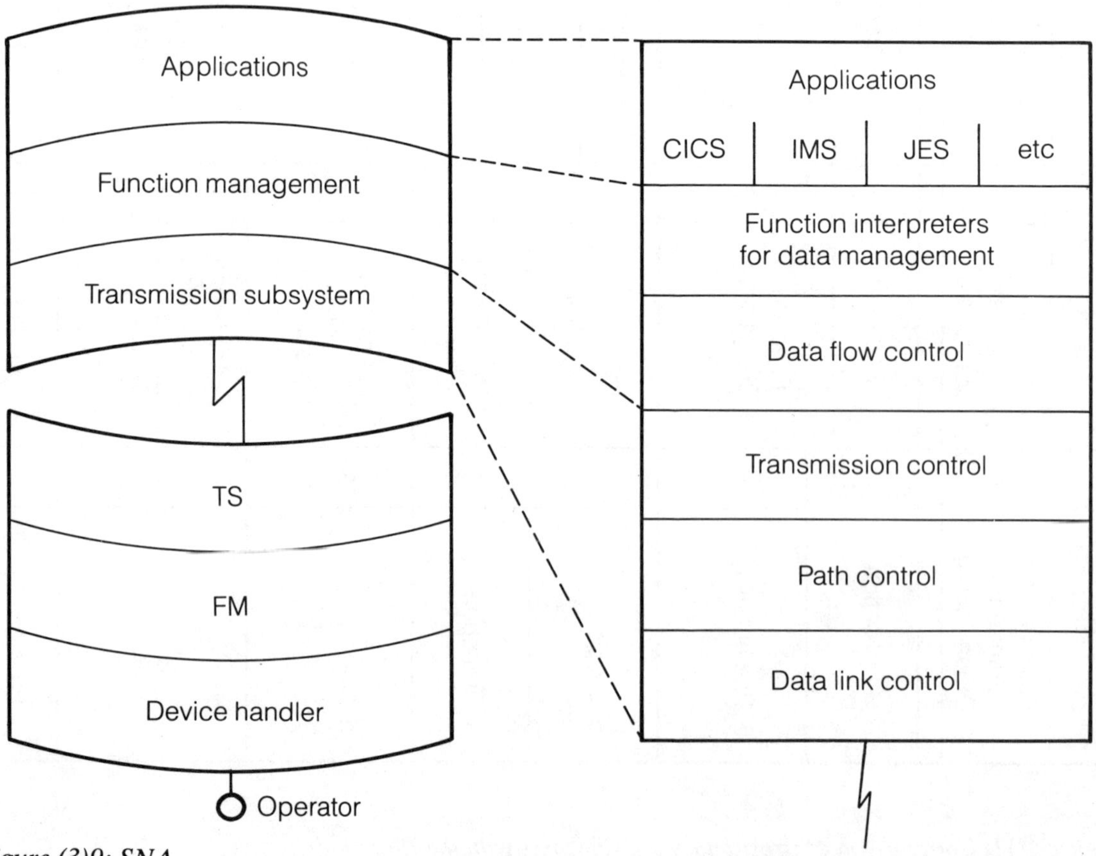

Figure (3)9: SNA

oriented facilities and functions that need to be provided in order to achieve a data communications environment, for data processing systems consisting of one or more host computers that would do the following:

1 Facilitate resource sharing....
2 Cater for efficient, effortless and unrestricted interactions between application programs....
3 Enable data communication functions to be distributed throughout the environment, not restricted to only the host computers in the system....

To illustrate the issues here, it is interesting to explore this point a little further. In an SNA network each terminal device is declared to the System Services Control Point (SSCP) which is within the VTAM software environment. A terminal is said to be within the 'domain' of its host SSCP. If the terminal users wish to access an application on another host within the SNA network, then the request is fielded by the domain SSCP and relayed to the SSCP in the target host. By agreement between the two SSCPs the routing tables in the front-end processors are modified so that the messages from the terminal to the target host, and *vice versa*, can be correctly routed (see Figure (3)10).

It will be noticed that this is a kind of switching and, of course, is quite consistent with the model for SNA which is a complete architecture that includes 'network' layer functions. It will also be noted that direct interworking between such a system and some other manufacturer's product will raise some problems because:

1 It requires that both have the same architecture.
2 Both must use the same message formats and protocols.

Proprietary architectures and interworking

For the sake of discussion, let us consider the possible ways in which another manufacturer's architecture product, DECNET for example, could be made to interwork with SNA.

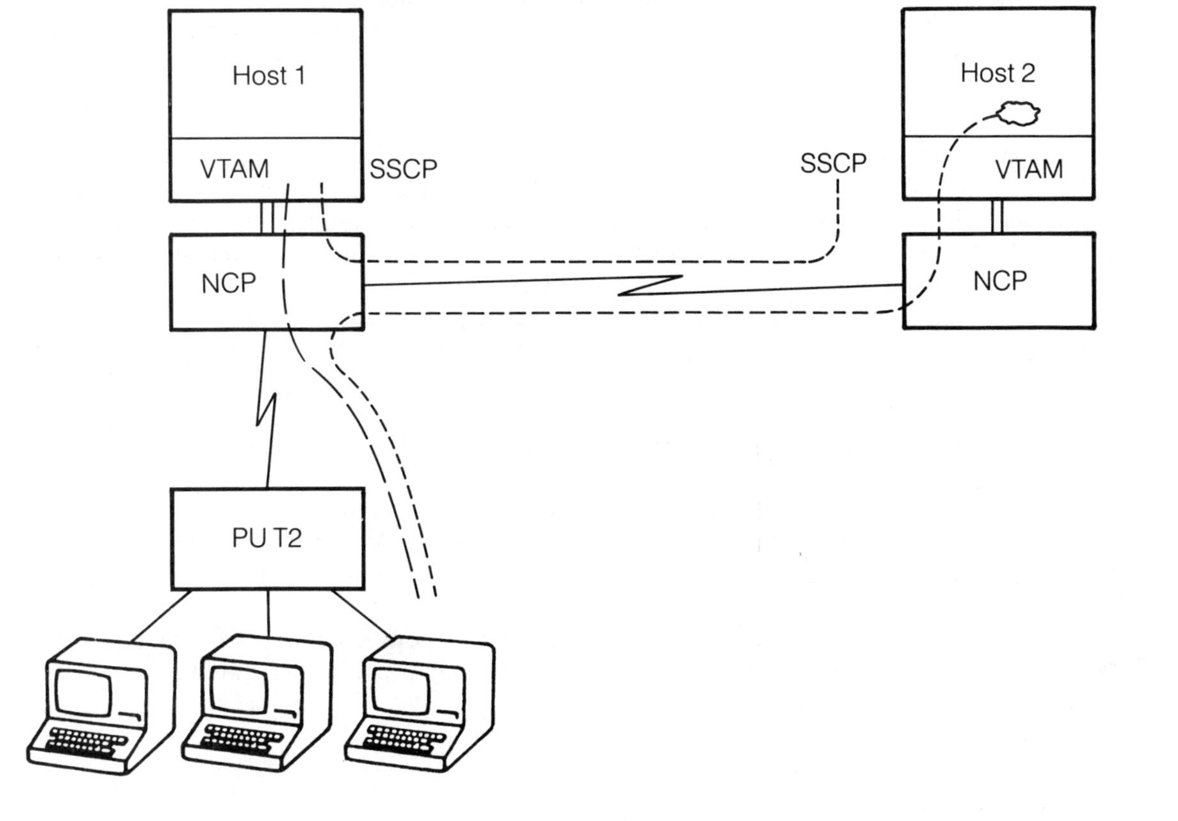

Figure (3)10: SNA ACF

The options would seem to be as follows:

1 Modify DNA to match SNA.
2 Map DNA onto SNA using a converter.
3 Use OSI.

Modifying DNA to match SNA

Modifying DNA (or any other manufacturer's architecture) to match SNA could mean adopting SNA as the *de facto* standard, but since much of DEC's software uses facilities in DNA/DECNET which may not be directly transferable to SNA, a considerable loss of investment is indicated.

Furthermore, it would require all the manufacturers to agree to the internal control, measurement, status message formats, SSCP procedures etc and for all these various implementations to be kept in step across various version releases. This degree of cooperation seems unlikely.

Before completely dismissing this approach, it must be remarked that IBM have announced a 'Logical Unit' type which, in principle, allows direct communication between intelligent devices (LU 6.2). In theory, this LU type could be emulated by other manufacturers' devices, thereby allowing them to communicate not only with an IBM host, but also with each other across an SNA network. Just how practicable or desirable this would be, remains to be seen.

Mapping DNA onto SNA by means of a converter

Many manufacturers already produce protocol converters which allow their products to appear to an SNA network in the guise of a known SNA terminal type. Many of these devices, however, are limited in the range of interconnection possibilities which they can support. For example, DEC produce a software package which allows a PDP-11 computer to appear to an IBM host as an IBM 3790 terminal system (Logical Unit type 2). Since this emulation program can run in the same software environment as DECNET, programs in other DEC computers linked to it, or terminals on the network, can access this emulation and invoke interaction with an IBM host computer (see Figure (3)11). Since the 3790 supports both remote job entry and 3270 datastream sessions, terminals such as DEC VT 100 can be given the appearance of an IBM 3270 transaction device and thus access program products such as CICS or IMS, while jobs can be submitted to JES applications on behalf of other users.

This type of product can be regarded as a 'gateway' between the two different environments. The term 'gateway' is used here in the same sense that it is used by the telecommunications network providers (ie between two otherwise sovereign networks). Not only does it convert the necessary signalling formats (protocols) used in the two networks, but it also acts as a point of demarcation across which network control and monitoring information does not pass.

In this case it is probably necessary to use the term 'architectural' or 'high-level' gateway in order to distinguish it from the communication network or 'low-level' type of gateway (see Figure (3)12).

The main disadvantages of this approach to interworking are as follows:

1 Limited functionality. Only certain classes of connection are supported and only, in the above example, from the DECNET side into SNA. Access from SNA terminals into DECNET is not provided, largely because the gateway is emulating a PU-2 which has no SSCP function.

2 For total interworking between any set of products, such a converter would be required between every pair of network architectures. Most manufacturers only provide such gateways between their own architecture and SNA.

Using OSI

Clearly if all manufacturers adopted a common OSI set of protocols, general interworking ought to be possible. Some manufacturers have indicated a policy to follow and adopt OSI protocols as and when they emerge. Others have not. However, as remarked above, a complete and satisfactory set of

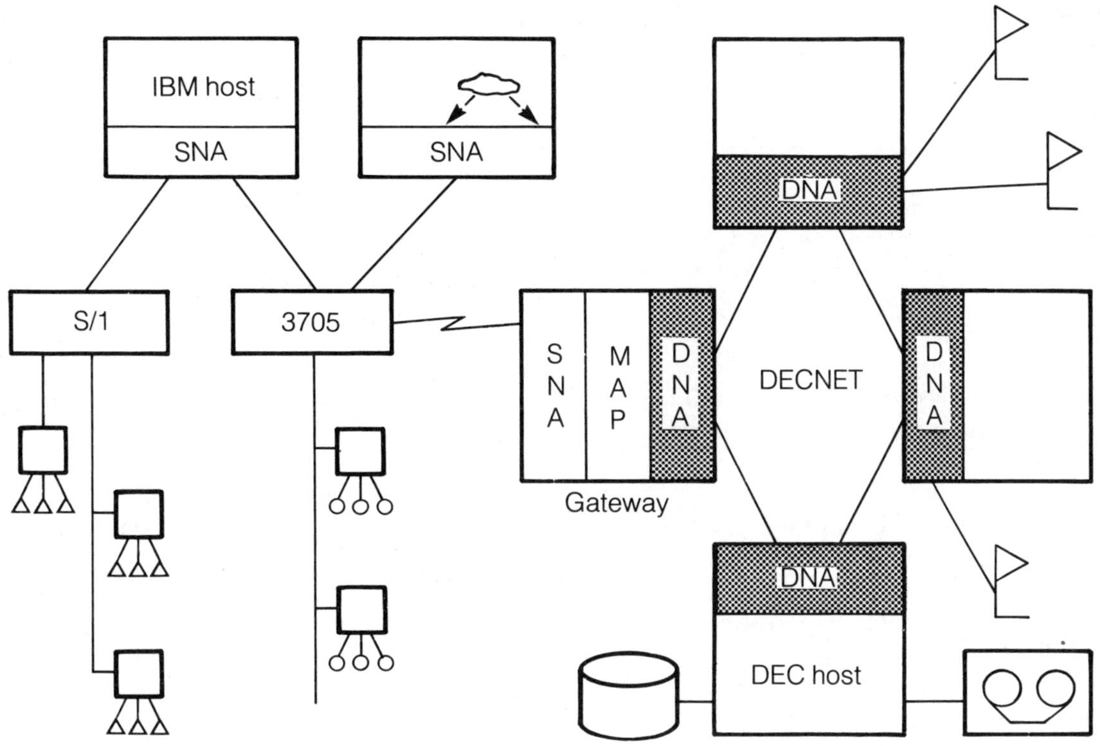

Figure (3)11: Architectural gateway

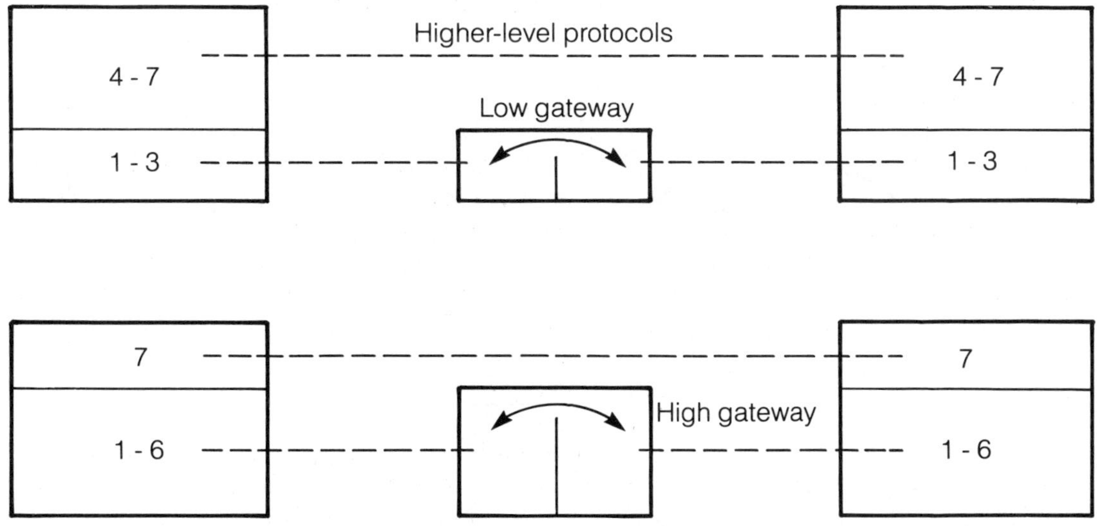

'Standard' half-gateways:

- ETHERNET, X.25, IP/TCP
- XNS
- Teletex, ODA
- SNA PUT2

Figure (3)12: Low and high-level gateways

protocols remains to be established and in the meanwhile the manufacturers must put something of their own in place. This means that products will have been produced and will require support even when the OSI protocols do finally appear (see below).

Tucker:
One can readily see the motivation for a computer manufacturer, wishing to get into OSI, to build an architecture which includes the stable parts of OSI together with some private 'best guess' or 'minimal function' implementation of the not-so-stable parts. Unfortunately such an apparently sensible migration strategy could, especially if adopted with different technical choices by many companies, lead to incompatible systems on the market which would all no doubt be claimed to be OSI in some sense. Furthermore, once such systems were built and sold, it would not be easy to upgrade them immediately to 'real OSI', however well intentioned the migration strategy might be. Thus early efforts to adopt OSI in practice could result in the addition of even more incompatibilities between not-quite-OSI and OSI, above and beyond those already permitted by OSI standards.

One possibility is for each manufacturer to produce a high-level gateway between the proprietary protocols and the OSI set. In theory at least, interworking is possible by interconnection of the various gateways (see Figure (3)13).

Whether this is practicable in performance terms and what range of facilities could be invoked is a very moot point. A more attractive approach would seem to be for manufacturers to provide the OSI set in parallel with their own. This would allow the continued use of parochial products within single-manufacturer networks, while the experience and functionality of OSI operation is being built up in a circumscribed and controlled manner (see Figures (3)14 and (3)15).

Tucker:
In whatever direction one might ideally like to see ISO or CCITT progress, one must accept that there are practical and political limits to their power to enforce compatibility through

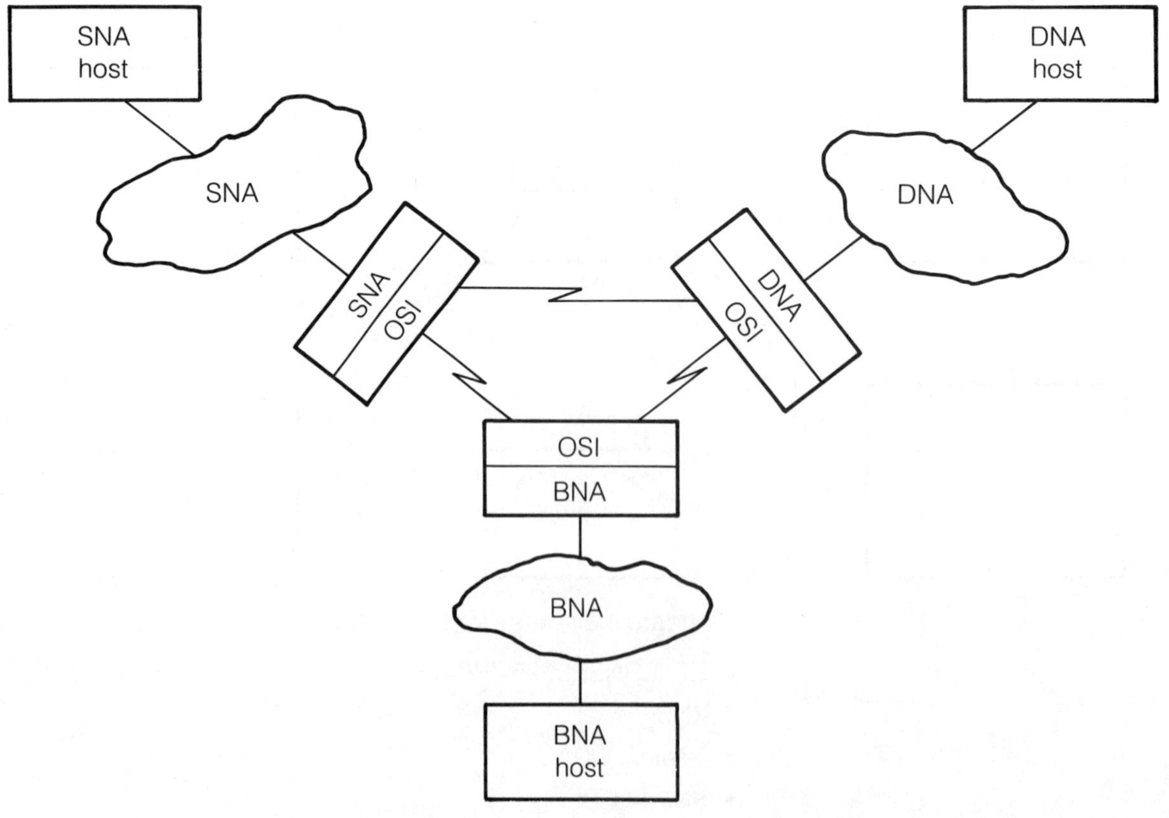

Figure (3)13: OSI gateway network

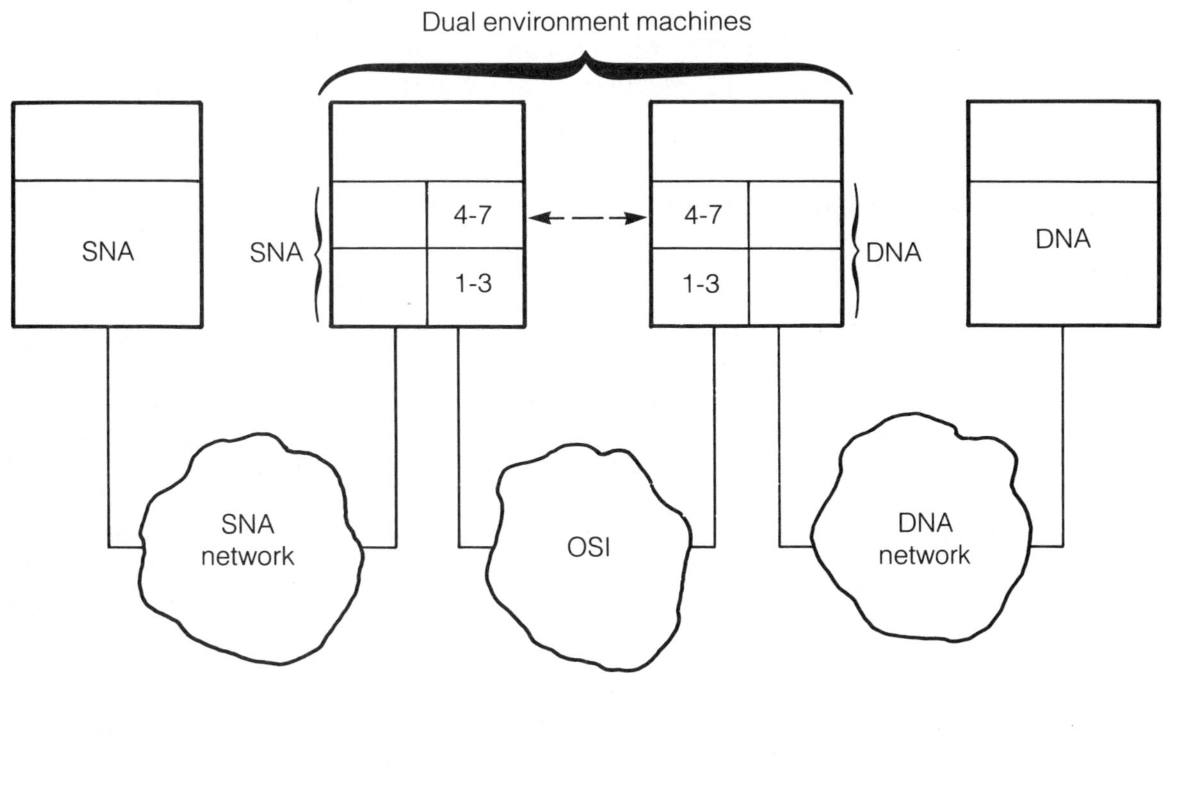

Figure (3)14: Interlacing OSI

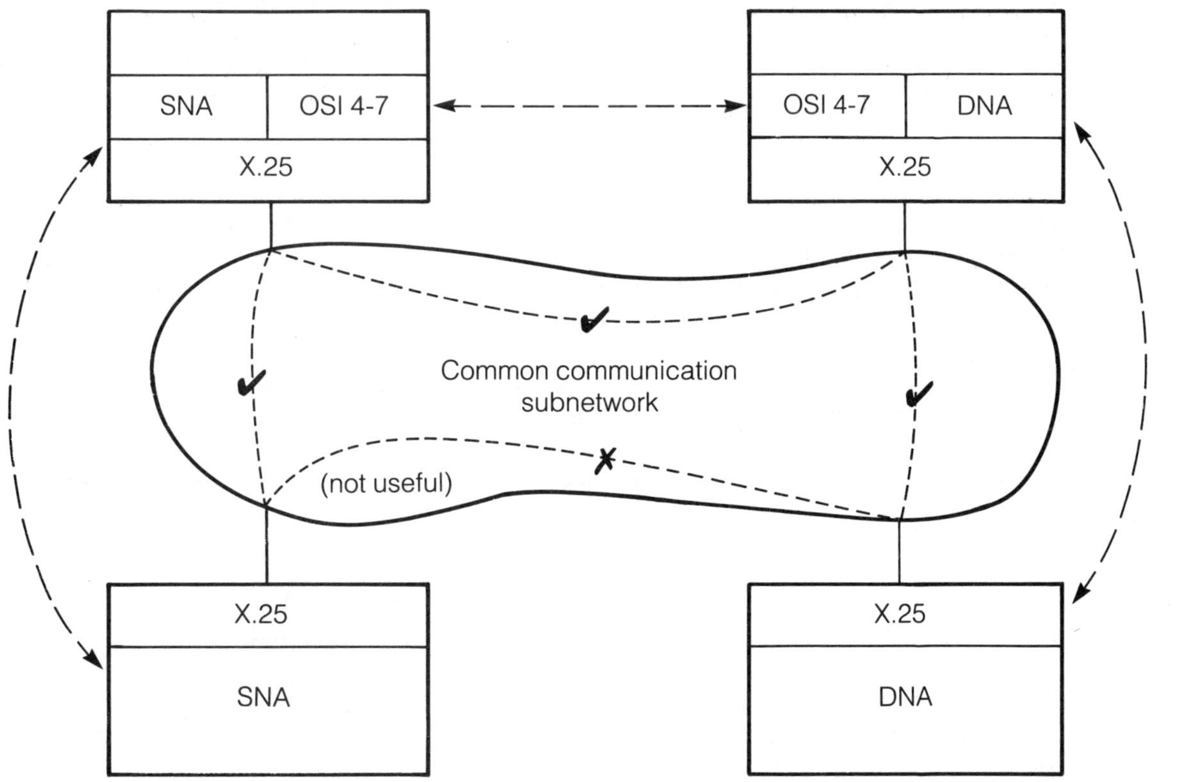

Figure (3)15: Common communication network

international standardisation. The final step to complete compatibility must be taken by others. Ultimately it is the market that will decide: if the standardisers cannot agree the 'conformance' protocols, natural selection in the competitive market will make the choice. Thus to a large extent the responsibility rests on those who commission or buy open systems: if they want guaranteed compatibility with other open systems, they must insist upon it — and be prepared to pay for it.

Management, operations and security

Network management

Most network products, be they basic transmission systems, switching systems or full-blown mainframe network systems, are provided with elements of network management and control. In more complex network systems, one or more special operator positions are provided to enable the system controller to access network management or network control functions. These functions are often concentrated in a network control centre or, as in the case of IBM SNA networks, in the host computer itself.

The degree of control exercised by the network management system depends on the nature of the product. In the case of a packet-switching communication subnetwork, the management and control will cover the switching nodes themselves and the interfaces to the telecommunications lines. (The latter may or may not include the modems or DCEs.) The control will probably not extend to the 'subscriber' equipment (DTEs), which may be both terminal devices and host computers. Contrast this with the view seen by the operators of a large host computer centre. From their perspective, the communication network is merely one of the links between the terminals and the applications, all of which fall under their purview. Add to this the specialist products such as the line and modem switching systems (which have their own monitoring and control systems) and it becomes apparent that there are overlapping and competing services. Moreover, these systems and services may or may not be under the control of one organisation. Imagine the scenario where the switched network is provided by a 'value-added carrier' (ie a VAN), the lines are provided by one or more PTTs, the mainframes by various suppliers and the terminals by a variety of users! (See Figure (3)16.) It is clear that considerable thought needs to be given as to how network management and control operations should be set up. In the past, the juxtaposition of these items has been very much on a case-by-case basis. However, with the advent of OSI, one may look for guidance as to how these functions could be integrated.

Tucker:
There is clearly a need for some functions that are not specific to particular communications but which are there to control and monitor the use of system resources, for example:

- *Updating routing tables or addressing directories*
- *Collecting statistics on traffic and performance*
- *Taking equipment out of use and reintroducing it.*

OSI management is still at an early stage of development and not unnaturally the highest priority has been accorded to developing the standards for 'normal' operation.

Such information needs to be passed off sideways and not up and down between functional layers (see Figure (3)17).

Sccurity

The term 'security' is used to cover a number of issues. These include the following items:

1 Continued operation in the presence of component or subsystem failure (availability).
2 Privacy of the information carried or processed.
3 Veracity of the information received.
4 Provenance of the information received.

The requirements for continued operation in the presence of failures can range from the very simple (it does not matter if it goes wrong occasionally, as long as it comes back quickly), to the very severe

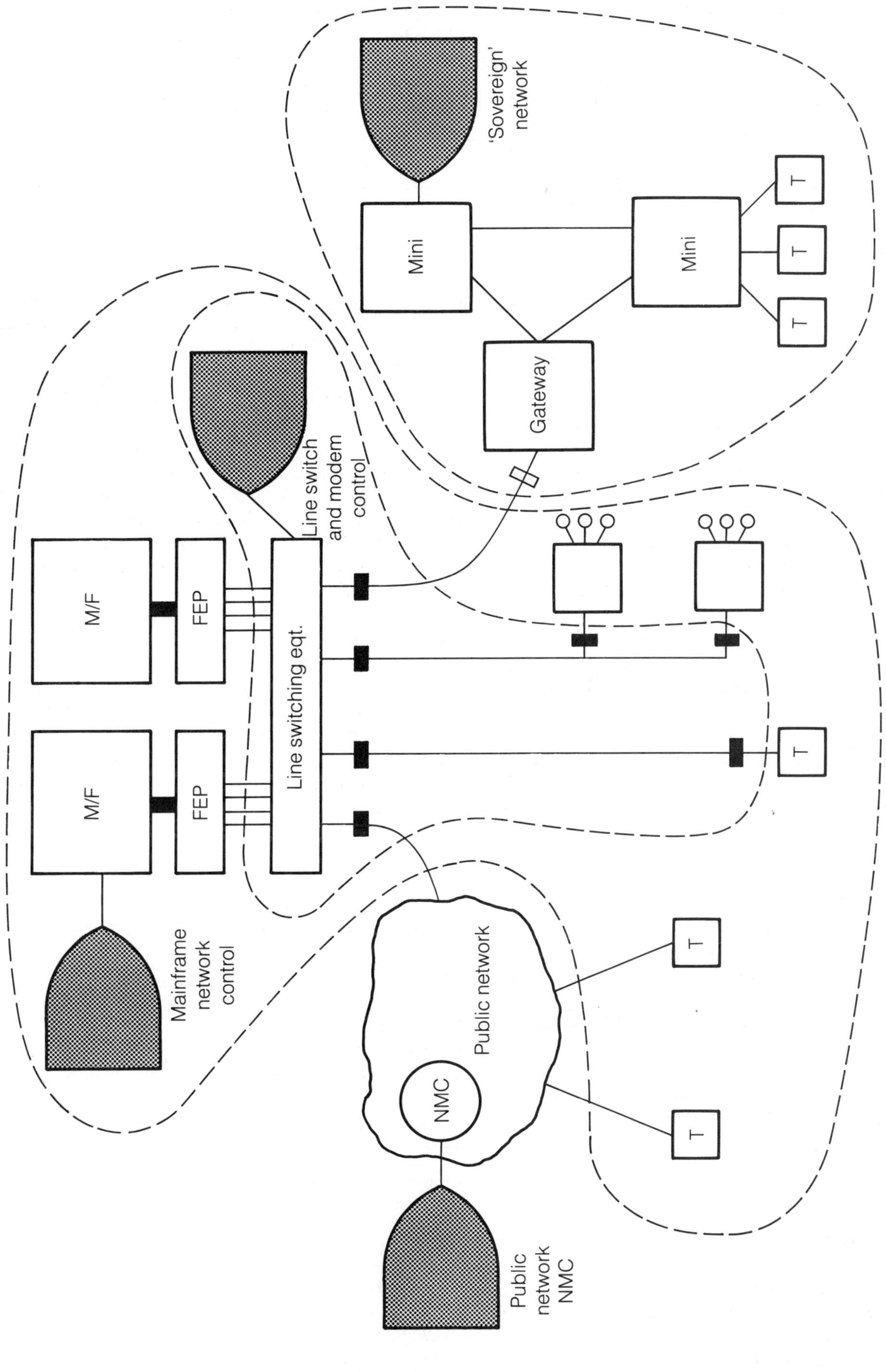

Figure (3)16: Network management and control problem

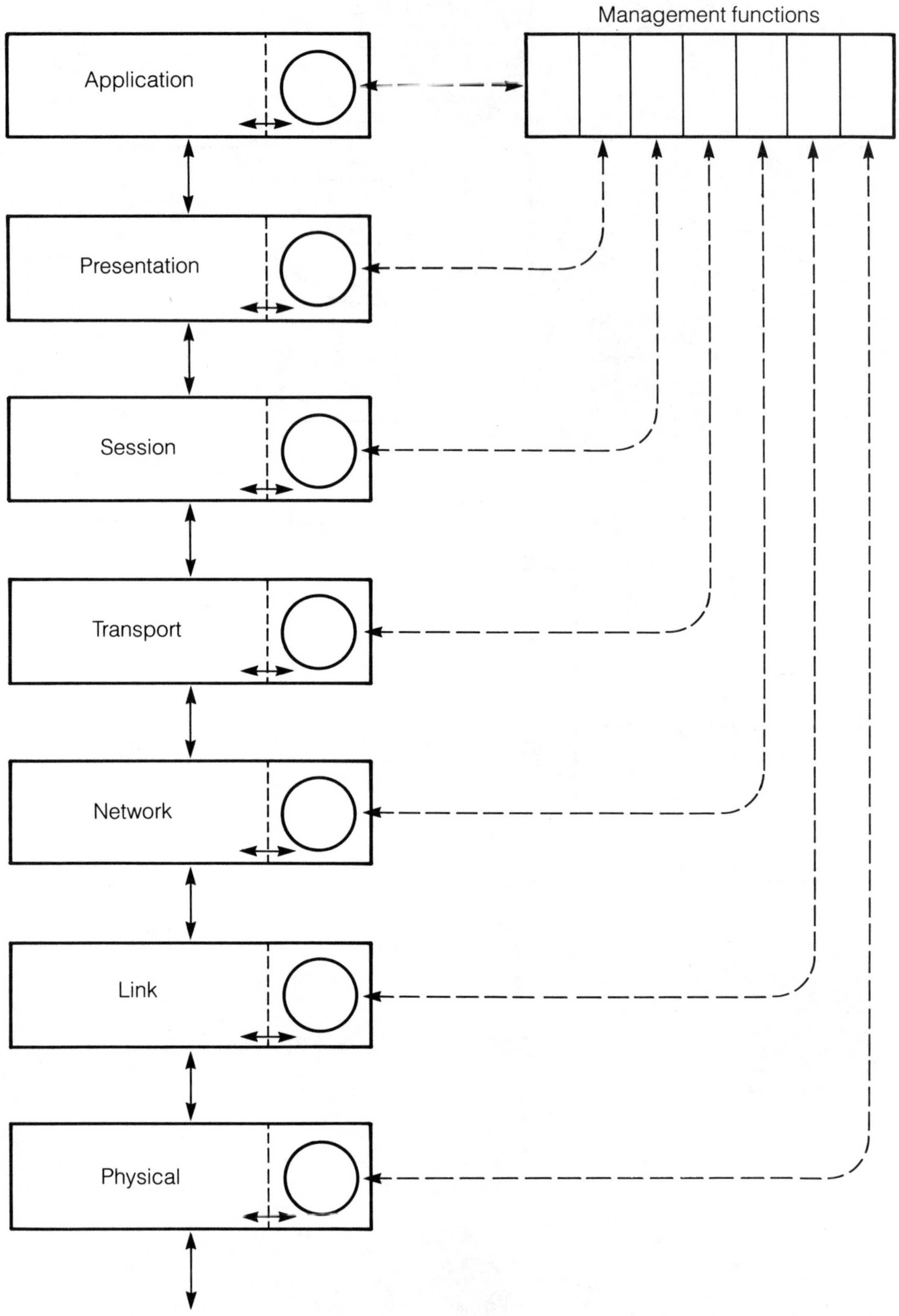

Figure (3)17: Sideways versus vertical movement of information

(under no circumstances must the system stop or any data be lost). In many instances, it is a case of providing an adequate degree of 'availability'. This is defined in the following way:

$$\text{Percentage availability} = \frac{\text{Mean time between failures}}{\text{Mean time between failures + mean time to repair}} \times 100\%$$

Thus high availability can be provided by increasing the MTBF and reducing the MTTR. One way of increasing MTBF is to provide extra equipment which is capable of taking over from failed equipment instantaneously, (so-called 'redundant' and 'hot standby' systems). A number of manufacturers provide these sort of systems (sometimes called 'non-stop', although this term is proprietary to the Tandem company). These systems can be quite expensive and if a lower grade of availability is tolerable, then simple duplicated systems ('warm' or 'cold' standby), or even just having a good stock of spare cards, may suffice. Having extra equipment present, of course, is a way of reducing the time to repair. However, replacement of failed equipment must be carried out quickly before more faults can occur.

The sensitivity to failure depends on the importance of the function. A central or common function failing may stop the whole system but a subscriber's line failing will only inconvenience one or a small number of users. The former may be deemed important enough to make special provision (eg hot standby). The latter may be deemed less significant since the affected user is merely inconvenienced (and could always go to another terminal unless, of course, he is a VIP).

Privacy of information is more difficult to achieve cheaply. Physical security can be provided in terms of preventing unauthorised persons entering locations where information is kept or where access to information can be made (eg from unsecured terminals). However, if information is to be processed or transmitted then it may be intercepted.

> *Price:*
> *We shall classify techniques designed to prevent disclosure of data under the heading 'secrecy'. Equally important is the need to prevent the undetected alteration of data that is being handled in a communication system. This property has been called 'authentication' of data; the term is used in this sense in some US national standards documents emanating from the banking community. Unfortunately, the term is also used within the OSI community in the sense of peer entity authentication, where communicating parties seek mutual verification of each other's identities; ... the term 'integrity' of data is now being adopted fairly widely to describe the process of preventing undetected alteration, reserving 'authentication' to mean peer entity verification. Complete prevention of alteration is frequently impossible; what is possible is to detect alteration when it happens.*

The techniques for providing these types of security (previously only found in military applications) are becoming available in the public domain. Again, with the advent of OSI, one may ask how these functions could be supported.

> *Price:*
> *Now that Open Systems Interconnection (OSI) has reached the stage where the architectural model is agreed upon, attention is concentrated on the development of communication protocols that fit within the architecture at the various levels. Among the services that may be included in these protocols are those designed to provide security for the data handled.... A fundamental technique in achieving secrecy, integrity and authentication is that of encipherment, where data is transformed by a complex algorithm under the control of an encipherment key. Within the last 10 years considerable progress has been made in the development of encipherment algorithms for use in the public domain ('public' as distinct from the world of military and diplomatic data protection). Publication of the US Data Encryption Standard (DES) has resulted in its subsequent adoption as an American national standard (known as the Data Encryption Algorithm (DEA); it is also being considered as a possible international standard.*

However, it should be realised that requirements for security, in any of its forms, need to be considered and specified at the outset. It is often impossible (or very expensive) to add these functions to an existing system that was not designed with them in mind.

4: Communication networks

The telephone has proved so important that the major national and international public networks are provided by regulated carriers, responsible to national governments and international agreements. The same is true for telex and international message-switching networks. The pattern is being repeated for the new data networks and for the move to the integration of these various services. On the other hand, private networks still abound. In the wide area network category, increasing pressure is being applied to these networks to make use of public switched services. However, within buildings and restricted areas, the so-called local area network technologies are springing up. The next important area confronting the networking fraternity is the issue of interworking between these rather different technologies.

Public data networks

The term 'public data network' can be thought of in terms of both switched and unswitched services. Thus in the UK we see the introduction of the new X-stream services from BT and the competing products from Mercury. There are also the burgeoning cable TV and Metropolitan Area Networks, although it is a little early to understand the effect these will have on the data network scene.

Burrows:
The licensing of the two rival systems for a set period is a good idea enabling developments to be monitored. It is likely that most large organisations will use both systems so that if one fails, the other system can be used. This trend will be set by the firms engaged in the financial sector and could also result in the decline in use of the leased line.... If the UK succeeds in becoming the first wired nation, it will have the following advantages:

1 It will have the most efficient communications network in the world, which will stimulate economic growth and investment.

2 There will be a major export market for this equipment.

The BT X-stream services (see Figure (4)1) can be regarded as providing a series of stepping stones from the old analogue telephone network-based 'datel' services towards the ISDN.

The introduction of the Packet-Switching Service (PSS) provided for the first time a public switched data network which was *ab initio* purpose-designed for computer traffic (see Figure (4)2). This service, and its counterparts in other countries, has had profound effects on the computer network scene, some of which are listed below:

1 Providing a private network involves an organisation not only with the cost of leased circuits and other communication equipment (multiplexors, etc) but also with the provision of network management and maintenance facilities. In a private network it is necessary to provide the logistics for detecting and recovering from equipment failures. It is also necessary to monitor traffic and anticipate growth requirements. The use of a public network removes the necessity for some of these activities. In the end it is a balance between the cost of undertaking these tasks as compared to the tariffs of the public or common carrier network. Many smaller users will find the public data network, like the public telephone network, an indispensable service.

Shields:
The usefulness of a service grows as the number of potential subscribers increases, and the ISDN provides the maximum connectivity for all users. However, this is only likely to happen if the ISDN can provide a satisfactory level of performance for data users at an economic cost.

- KiloStream 2.4 Kbit/sec
  4.8 Kbit/sec
  9.6 Kbit/sec
  48 Kbit/sec
  64 Kbit/sec

- MegaStream

  MegaStream 2 (2 Mbit/ or 30 x 64 Kbit/sec)
  MegaStream 8 8 Mbit/sec

- SwitchStream

  SwitchStream 1 PSS
  SwitchStream 2 ISDN (System X)

- SatStream

Figure (4)1: BT X-Stream services

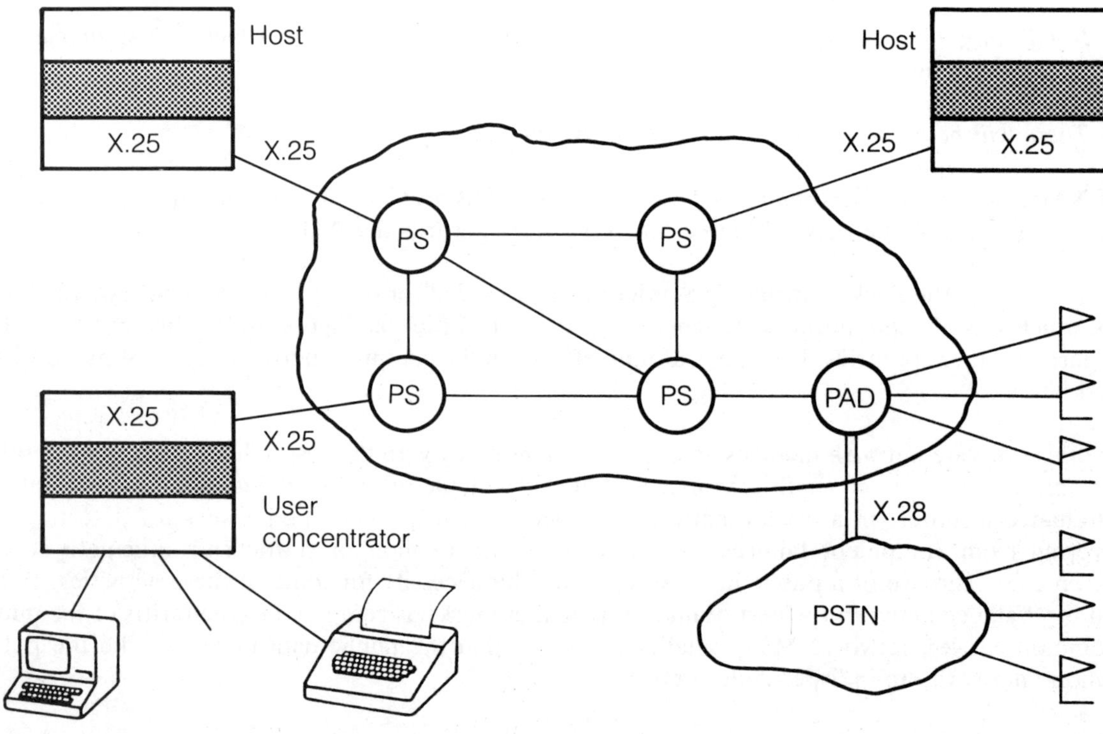

Figure (4)2: Public data network (packet)

2 Many large organisations have private data networks but cannot, of course, use them for communicating with customers or suppliers that do not have access to the network. An individual, on the other hand, may need access to several such networks and it is not generally economic in this case to provide multiple connections and multiple terminals. (Take for example the requirement for a retailer to carry out authorisations on a variety of credit cards.) In this case the common use of a public network is indicated.

3 The introduction of the X.25 interface for packet-switching networks has arisen from the recognition of the need for an interface suited to certain kinds of computer traffic (rather than making do with adaptations for using the telephone network) *(015)*. The widespread introduction of X.25 services has encouraged manufacturers to support the standard and users to press suppliers and carriers alike to provide interworking capabilities on this basis (in particular the international interconnection of national X.25 networks).

4 Penetration into the domestic and small office market depends on cost. Most buildings do not have data network terminations, but do have the telephone. Access to PSS has been extended by the provision of dial-up facilities for simple terminal devices (the so-called Packet-Assembly-Disassembly service or PAD). The interworking of the two networks and the provision of the added services presages the ISDN.

Some other national communication administrations have seen fit to introduce public data networks based on circuit-switched technology using the X.21 interface instead of X.25 (see Figure (4)3). This has happened in the Nordic countries (Denmark, Finland, Norway and Sweden) where the Nordic Public Data Network has been in operation since 1981.

Although the network is being used by a number of subscribers, there is also pressure for the administrations to provide an X.25 service so as to permit interworking with other countries; there being more X.25 networks presently in operation (see Figure (4)4).

The Federal Republic of Germany also started with the provision of a circuit-switched public network (Datex-L) but has already provided a packet-switching service as well (Datex-P).

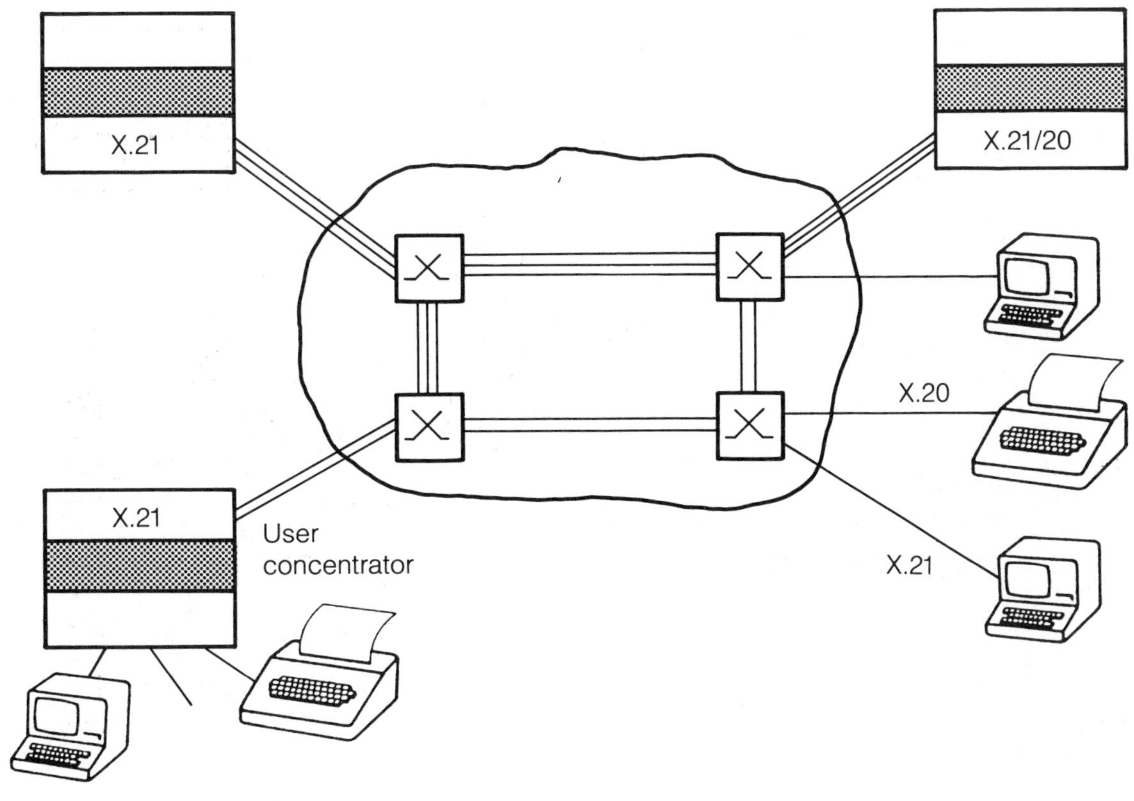

Figure (4)3: Public data network (circuit)

a)

Country	X	Network
USA		Tymnet
		Telenet
Canada		Datapac (TCTS)
	X	Infoswitch (CNCP)
		Teleglobe (COTC)
Belgium		DCS
Finland	X	Datapak
Denmark	X	Datapak
France	X	Transpac
W.Germany	X	Datex-P
Italy		Itapak (?)
Netherlands		DNI
Norway	X	Datapak

b)

Country	X	Network
Portugal		Telepac
Spain		Iberpac
Sweden	X	Telepak
Switzerland		Telepac
UK		IPSS
		PSS
Japan		NTT DDX-2
		KDD Venus
Australia		(Austpac + OTC)
Singapore		Telepac
Indonesia		Packsatnet
S. Africa		Saponet

c)

Access to Telenet or Tymnet

USA	Israel	Spain
Argentina	Italy	Sweden
Australia	Kuwait	Switzerland
Austria	Japan	Taiwan
Bahrain	Mexico	UK
Belgium	Netherlands	
Bermuda	New Zealand	Alaska
Canada	Norway	Hawaii
Denmark	Philippines	
Dubai	Portugal	
Finland	Puerto Rico	
France	Qatar	
W.Germany	Saudi	
Hong Kong	Singapore	

Key: X = Also circuit switch X.21

Figure (4)4: Countries with X.25 networks

At the moment the pendulum seems to be swinging in the direction of X.25 and packet-switching, but it remains to be seen how this will be affected by the introduction of ISDN.

However, it must be remembered that the fact that two computers use X.25 does not necessarily mean that they can interwork. It is still necessary to agree the common use of higher level protocols. Perhaps the opportunity exists for a public protocol conversion service as shown in Figure (4)5.

ISDN

As mentioned elsewhere, the ISDN is designed to provide a single communications network environment giving access to switched and non-switched services including voice and data and possibly other facilities as well; all of this being achieved over a single customer connection and termination mechanism (IDA).

> *Burrows:*
> *There are two aspects in the technical development in communication networks. Firstly, new methods such as the use of fibre optics, area networks and teleconferencing are developing at an increasing pace and thus creating new services. At the same time, these services are converging into a common technology by the use of digital transmission. For example, both voice and data can be transmitted over the same network and, by the use of digital technology, facsimile is being integrated into this system. This means that graphics could soon be integrated into the same system as voice and data communications.*

> *Ettinger (1):*
> *The introduction of the digital, Time Division Multiplex (TDM) PCM techniques in subscriber networks will not only lead to the lowering of costs but may also open up many new service options. The twisted pair can, in principle, carry data at rates of up to 144 Kbit/sec. This allows for the introduction of digital services such as teletex, videotex as well as voice, right down to the subscriber terminal equipment. These services may also include data transmission, electronic*

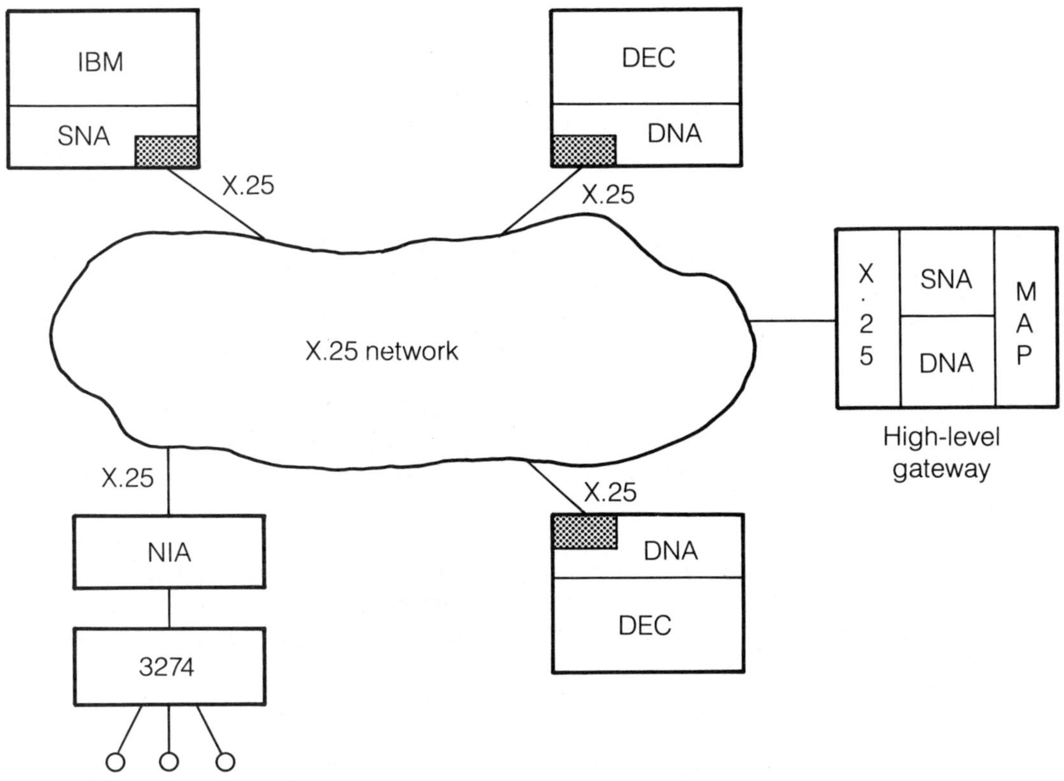

Figure (4)5: Public network support

mail, teleconferencing and electronic fund transfer, which will be carried in accordance with a verified set of digital standards, within an Integrated Services Digital Network (ISDN).

One of the major problems in introducing new facilities or technology into the public network is that it cannot be done overnight. This means that whatever is introduced must coexist and interwork with the existing system, ie evolution not revolution.

For example, in the new system the voice signals will be digitised and transmitted at 64 Kbit/sec. New digital switching systems are being installed (System X) which will switch the digital signals in their digital form *(016)*. However, as there are over 6000 telephone exchanges in the UK it will be some time before all the analogue plant and old step-by-step signalling has been replaced. Not until this has happened can the voice path be universally digital from end to end.

Ettinger (2):
The programme to replace current analogue exchanges with digital exchanges is likely to be long term. However, there may be potential customers for ISDN in areas not likely to be serviced by a digital LE in the foreseeable future and in such cases IDN multiplexors will be used to provide ISDN services ...

Any business introducing new technology and facilities wishes to do this in a smooth planned manner with minimum disruption to current work. This process is sometimes referred to as 'migration'. In a medium to large business organisation, consideration will need to be given not only to the types of application, but also to the geographical locations of branches and subsidiaries. Furthermore, it may be uneconomical to write off all the existing equipment at the same time and replace it by new (even if a PTT should be capable of providing the new facilities simultaneously at all the relevant locations). It is therefore an important facet of ISDN to provide full facilities for analogue ports to which the user's current analogue equipment may be connected eg telephones, analogue PBX or modems.

Parts of the network will be cut over to digital operation while other parts are still analogue. This means that analogue to digital signal conversions will be used at the required points. Meanwhile interworking between data terminals will still require the use of modems where either of them is connected to the analogue network.

In anticipation of the digital switching, a significant amount of digital transmission plant has been installed by BT and some of this 'digital overlay network' is being offered for point-to-point data transmission purposes as the above-mentioned Kilostream and Megastream services. These services are starting to prove to be very attractive in terms of their cost/performance in comparison with the older analogue-based services.

BT are also conducting a pilot trial of the Integrated Digital Access service which will allow the customer's digital termination to be shared between the telephone and other data or non-voice applications. Remarkably, the proposed interface arrangements for connecting to the network are different from those already in use, either for the telephone or for data networks.

In ISDN/IDA, a separate channel is used for network signalling (dialling the call or receiving the ringing tone), whereas in most existing networks such signalling is sent over the same interface as is used for subscriber-to-subscriber signals. Presumably, therefore, adaptor boxes will be provided for handling X.21 or X.25 terminals!

This raises another interesting issue which concerns the nature of ISDN services. The concept is based on circuit-switched calls at 64 Kbit/sec (ie the voice requirement). How, then, will X.25 devices be accommodated, and what facilities will exist for users or services requiring data rates higher than 64 Kbit/sec?

Burren:
One has the feeling that too heavy building in of 64 Kbit/sec would be a mistake. Computer people will see warning parallels with computer architectures having address ranges of 64 Kbytes! Provision of broadband channels is envisaged in the ISDN proposals and is currently under study. One hopes that recommendations will be agreed in the near future.

The view may be taken that it is quite possible to provide access to such services (eg PSS) using the IDA service. Just how this is accomplished technically does not matter to the subscriber, provided the service parameters are suitable (and hopefully unchanged) (see Figure (4)6).

> *Ettinger (2):*
> *Although the user will no doubt be influenced by the tariff structure he must have the final choice of which services he might wish to use.*
>
> *The basic conceptual principles of ISDN as outlined in I-Series Recommendations are that:*
> - *It should evolve from the telephony IDN based upon 64 Kbit/sec channels*
> - *It should have both circuit- and packet-switching capabilities from a family of digital user/network interfaces*
> - *Where applicable, it should possess hybrid access capabilities.*

What it does mean is that inside ISDN a number of different services will coexist. The integration will be in terms of equipment rather than services, and this process starts in the subscriber's premises where the cable comes through the wall or onto the desk.

> *Ettinger (2):*
> *It must be recognised that for business users integration between communication and office automation and between data, voice, text and image should start right in the 'business area' ...*

Text and information systems

The usefulness of telex cannot be denied and the increasing use of communicating word processors, mailbox services like BT Gold, etc indicates that the need for standards in the text and information area is just as urgent as in the data processing area.

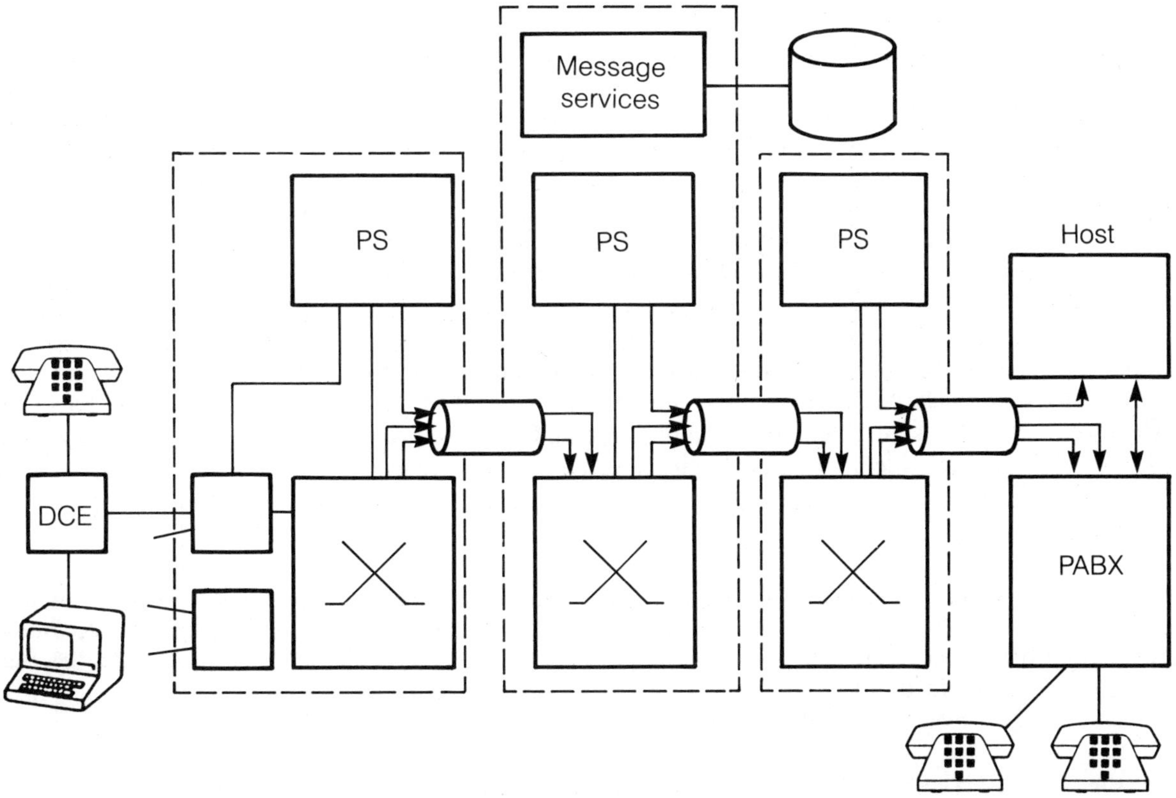

Figure (4)6: ISDN

Apart from the manufacturers' products, such as IBM's DISOS and Wang's Mailway, a number of public-domain standards are emerging such as:

- Teletex
- Viewdata/videotex
- Message Handling System architecture (MHS)
- Message Interchange and Document Architecture Standards (MIDAS).

All of these, with the exception of MIDAS, are CCITT/PTT inspired. The MIDAS proposals come from the ECMA stable and might be regarded as being in conflict with the CCITT proposals.

> *Willmott:*
> *The main current difference is that whereas the CCITT considers only the interconnection of public administrations messaging systems, ECMA and ISO are also concerned with the interconnection of private messaging systems and connections between private and public administration systems. Hopefully all their efforts will result in a single coherent set of standards for general application.*

Teletex

Teletex is designed as the replacement for telex *(017)*. It provides machine-to-machine message capability at 2400 bit/sec using an extension of the International Alphabet number 5 (see Figure (4)7). The service and protocol architectures are given in the CCITT recommendations shown in Figure (4)8. The architecture is layered in the approved fashion but the layers and the protocols match the OSI model less well towards the top end. In the UK it is assumed that the service will initially use the public telephone network and that terminals can be obtained from any supplier. Indeed, as the service is designed for memory-to-memory operation with subsequent printing off-line (ie not interactive) it is also assumed that word processors or other computers could equally well interwork in this mode.

Other administrations take a different view. In the Nordic area the use of the public data network is required and the terminals will be supplied by the PTT. Interworking between Teletex terminals on an international basis will therefore require the provision, in some instances, of special Teletex gateways in order to connect up the different networks. Limited interworking is also foreseen between the telex and Teletex networks by means of gateways providing appropriate format conversion services. Strangely, the current ISDN proposals do not appear to dwell on these particular issues.

> *Ettinger (2):*
> *Teletex terminals will be supported by the IDN either by use of a modem connected to the analogue port or by means of synchronous digital interfaces. Modem-based terminals will be capable of interworking with each other by using ordinary telephone calls without any interworking units. For interworking with packet terminals the PNA will be needed, while for interworking with telex the Teletex conversion facility as shown* [Ettinger (2), Figure 13] *will be required.*

A paper published in IBM's System Journal gives a cost comparison for Teletex against telex and letter post in West Germany in 1981. Transmission of a four-page document by Teletex would have cost between six and 19 US cents, depending on distance and time of day, compared with 89 cents to $4 for a telex and 94 cents for a letter (including the internal handling costs).

Viewdata/videotex

The original Prestel system from BT was seen as a service for both the domestic and business market *(018)*. As such, the use of the domestic colour TV receiver and the common coding scheme for dial-up viewdata and broadcast teletext services such as Ceefax and Oracle (not to be confused with Teletex), was a good idea. Unfortunately the cost of telephone calls and the limited choice of material in the database initially proved unattractive to the domestic user.

After this false start, viewdata services are now proving popular in the commercial market (eg the TOPIC service supplied by the London Stock Exchange) *(019)*. Plans for the Prestel service in UK and the Bildschirmtext service in Germany are illustrated in Figures (4)9 and (4)10. Again the problems of wider penetration are bound up with the issue of standards. Other countries, notably Canada and

Primary graphics: normal roman printing characters (columns 2–7)
Supplementary graphic characters (columns 10–15); Non-spacing diacritical marks (column 12)
Primary control codes (columns 0–1)
Supplementary control codes (columns 8–9)

	0	1	2	3	4	5	6	7	8	9	10	11	12	13	14	15
0	·	·	SP	Ø	@	P		p	·	·		°			Ω	k
1	·	·	!	1	A	Q	a	q	·	·	¡	±	`		Æ	æ
2	·	·	"	2	B	R	b	r	·	·	¢	²	´		Đ	đ
3	·	·		3	C	S	c	s	·	·	£	³	^		ª	ð
4	·	·		4	D	T	d	t	·	·	$	×	~		Ħ	ħ
5	·	·	%	5	E	U	e	u	·	·	¥	μ	¯			L
6	·	·	&	6	F	V	f	v	·	·	#	¶	˘		Ĳ	ĳ
7	·	·	'	7	G	W	g	w	·	·	§	·	˙		Ŀ	ŀ
8	BS	·	(	8	H	X	h	x	·	·	¤	÷	¨		Ł	ł
9	·	·	)	9	I	Y	i	y	·	·			¨		Ø	ø
10	LF	SUB	*	:	J	Z	j	z	·	·			˚		Œ	œ
11	·	ESC	+	;	K	[	k		PLD	CSI	<<	>>	¸		º	ß
12	FF	·	,	<	L		l	|	PLU	·		¼	_		Þ	þ
13	CR	·	–	=	"	]	M		RLF	·		½	˝		Ŧ	ŧ
14	·	·	.	>	N		n		·	·		¾	˛		Ŋ	ŋ
15	·	·	/	?	O	—	o	·	·	·		¿	ˇ		'n	

Figure (4)7: Teletex code table

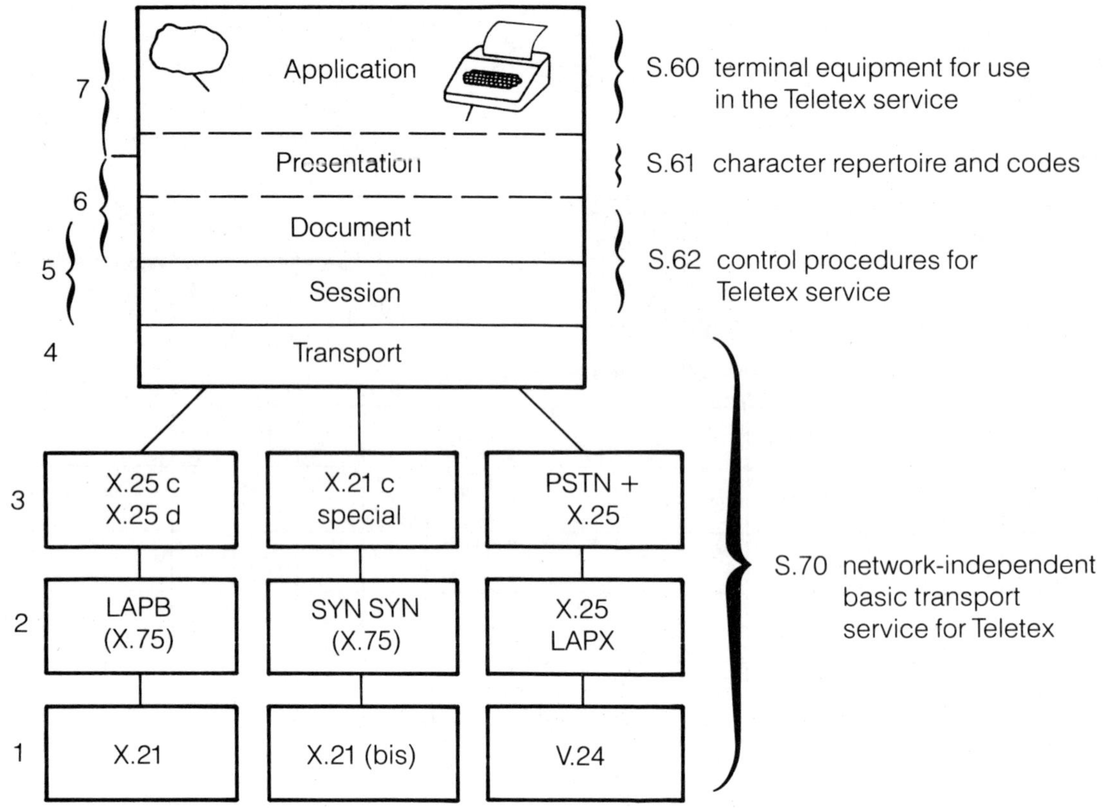

Figure (4)8: Teletex architecture (F.200)

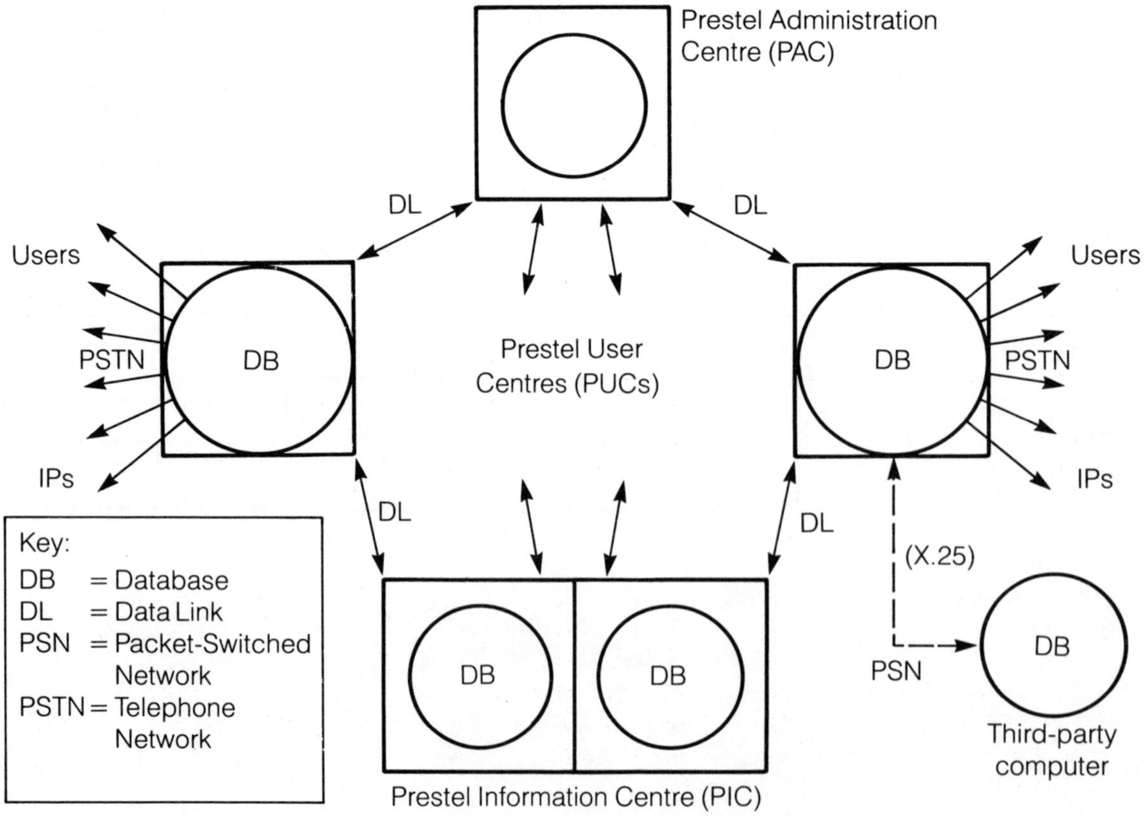

Figure (4)9: Prestel Advanced Network Design Architecture (PANDA)

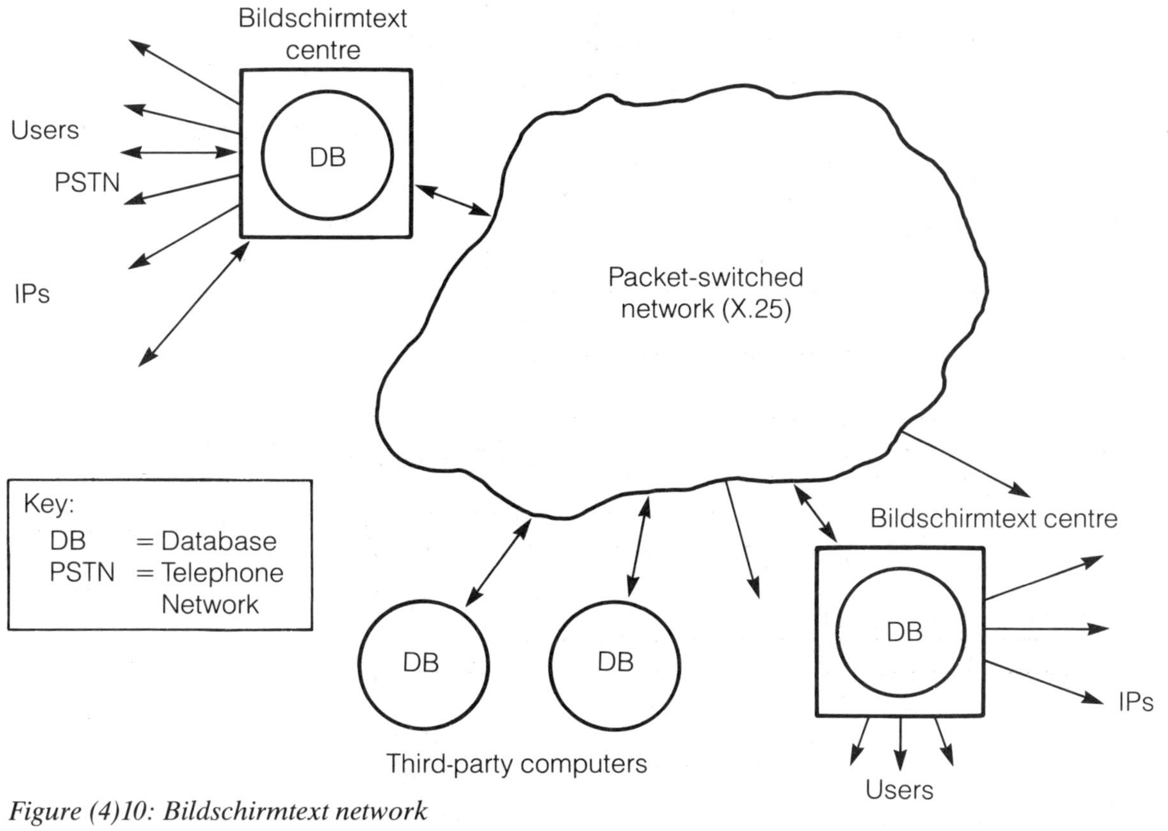

Figure (4)10: Bildschirmtext network

France have put forward different systems. The generic term for viewdata and Teletext is now 'videotex' and common coding schemes are being sought in CCITT *(020)*.

Message Handling Systems (MHSs)

More recently, significant amounts of work have been put into the CCITT MHS recommendations (X.400 series). These in effect create a substructure within the Application Support Layer (Layer 7) of OSI, and are specifically designed to define the text and message handling environment.

The competition is essentially with the products of the computer and word processor manufacturers. Again the solution might be to provide both the parochial product (for intrasystem working) and support for the standard (for intersystem working).

> *Willmott:*
> *This set of standard services and protocols for store-and-forward message transfer (the X.400 draft recommendations) provide a flexible means of interconnecting many diverse messaging systems, CBMSs, message-switching systems and, in particular, standard ways of connecting the telex/teletex community to the messaging community. CBMSs and gateways will be developed to allow other isolated communities such as videotex, word processors and personal computers to access the messaging community.... Such a solution is not readily applicable in the case of more complex terminals which might be capable of combining character coded text, fax and other types into a single complex message. This is a case of associating the message content with a specification of message body content type, logical and layout descriptions. These descriptions are forwarded with the message content to enable decoding of the message content at the recipient's terminal. This clearly requires a high degree of intelligence at the recipient's terminal and is therefore unsuited to the simpler terminals.*
>
> *Standards for this type of message content description are currently being developed by CCITT (T.73); ECMA (office document interchange format) and ISO (office document architectures).*

Local Area Networks (LANs)

In telephone systems the local network has tended to be taken for granted by the users. It has been viewed perhaps as a natural extension of the public network. Even where it has been necessary to install a PABX, it has never been seen as much of an issue and certainly not a controversial one.

Perhaps this is why there are two camps when it comes to discussing the topic of LANs:

1 The combined voice and data PABX (see Figure (4)11).
2 The separate data LAN (see Figure (4)12).

The PABX topic has been previously discussed. Why, then, should the separate data LAN be discussed when considering integration? The simple answer is that voice and non-voice applications seem to have radically different traffic patterns and this has led to the development of specific technologies for dealing with them.

The data LAN offers two potentials which can be considered as supporting integration, namely:

1 A common medium for connecting up data devices within a building (or limited geographical area).
2 A high-speed trunk system which can be used like an extended computer bus.

The first point can, of course, also be met by the PABX, subject to certain limitations like data rate. The second point, however, is presently only found in the high-speed dedicated data LANs (ie those operating in the range of 1 Mbit/sec and above).

This mode of use, though, is only invoked in those systems or product lines which have been designed for 'close-coupled' operation. Typically this might be 'office automation' products where physically distributed functional 'servers' and intelligent workstations can exchange data blocks at high speed.

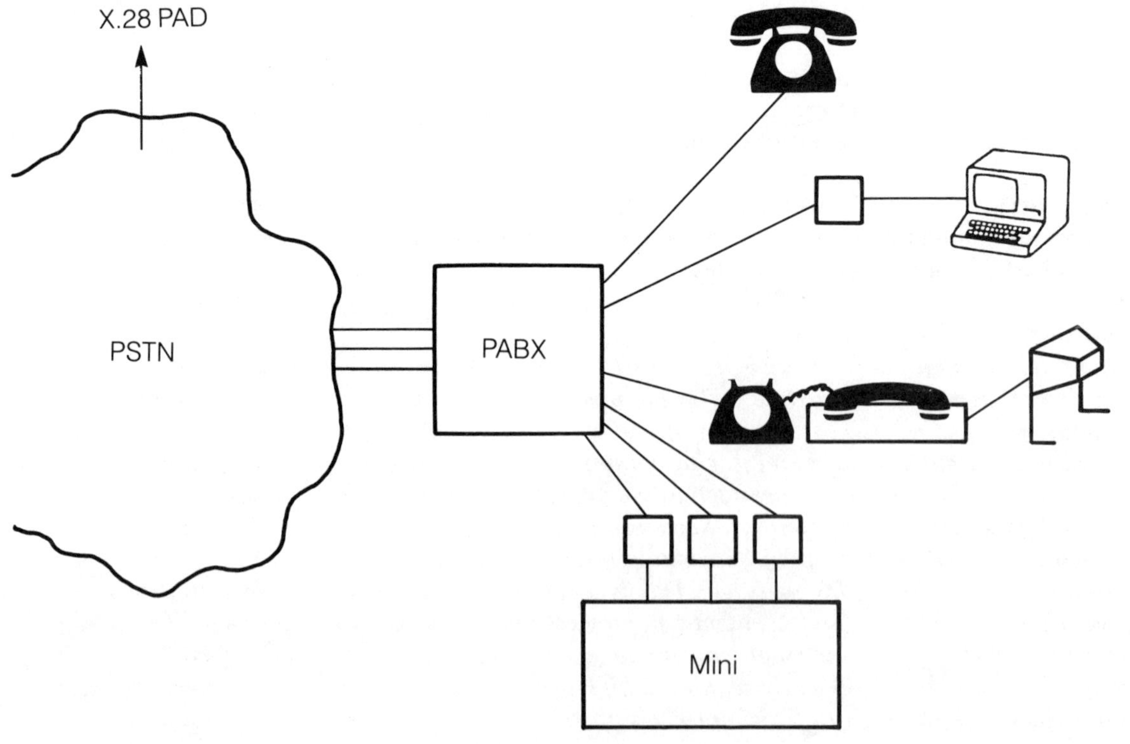

Figure (4)11: PABX-based LANs

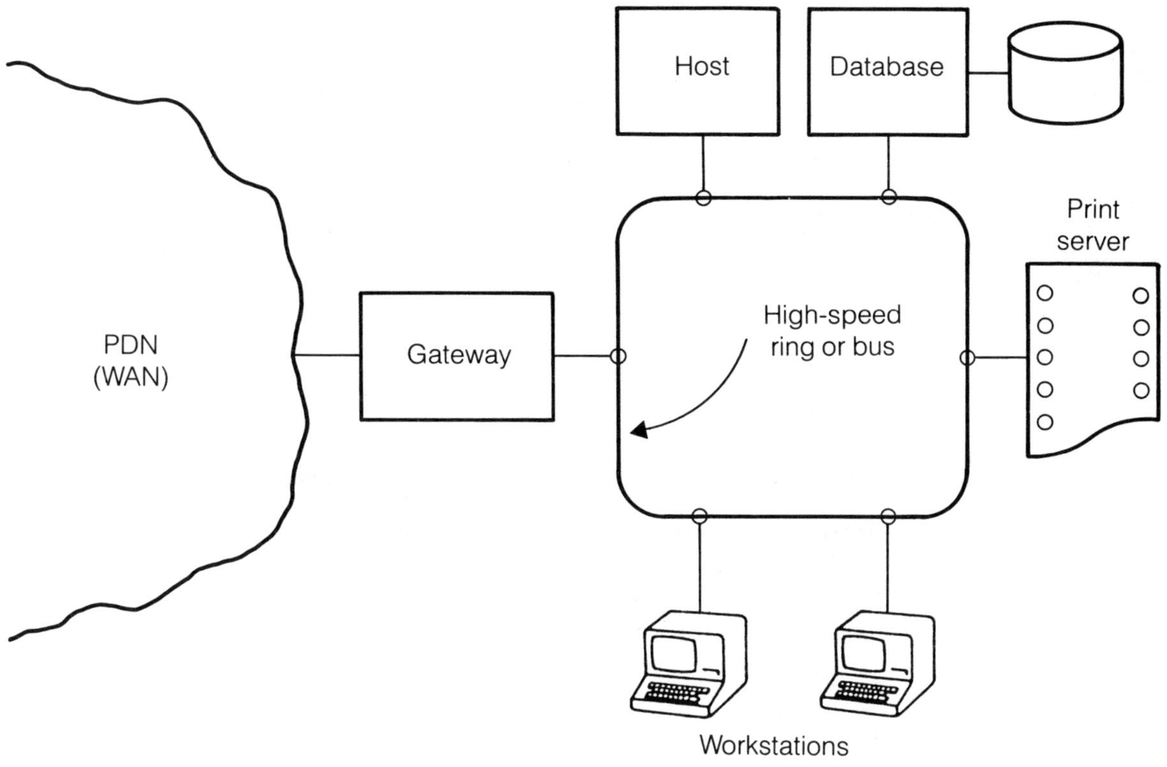

Figure (4)12: Specialised LANs

Unfortunately close-coupled operation also means that specially designed interfaces to the LAN are necessary together with agreed protocols and formats for interworking and a distributed operating system concept. This point of course raises the whole issue of standards and architecture all over again, but in the context of LANs.

> *Burren:*
> *... the ISO model and the ISO standards have become a formidable driving force in the development of computer systems. There are, however, some difficulties with the model. Hierarchical structures have the general difficulty that they assign a fixed ordering of the functions and this ordering may be unsuitable to changes in application or technology. There have already been difficulties in accommodating LAN and satellite transmission technologies within the model.*
>
> *A second difficulty is that the model and standards do not take account of speed.*

LAN standards

The race between products and standards in LANs mirrors the history of Wide Area Networks (WANs). At present there is little or no work on LANs taking place in ISO or CCITT and the field has been left to bodies like the IEEE in the US and ECMA in Europe.

The IEEE project 802 is now quite well known and an analysis of how the present structure of work was formulated is quite illuminating (see *(021)* and Figure (4)13). Suffice it to say the work had to bow to the inevitable and take into consideration the momentum behind the Ethernet standard as well as the views of manufacturers not wedded to this approach (eg IBM's support for the token ring approach).

On the other hand, the IEEE, being an American body, has not been able to take into consideration the developments occurring in Europe, for example the not inconsiderable community in the UK using the 'Cambridge Ring' technology *(022,023)*.

High levels 802.1

Logical link control 802.2

CSMA/CD (ETHERNET) 802.3

Token bus 802.4

Token ring 802.5

Metropolitan area nets 802.6

Technical Advisory Group 1: Broadband techniques

Technical Advisory Group 2: Optical fibres

Figure (4)13: IEEE Project 802

Although much ingenuity has been shown in the various techniques for improving signals on LAN cables, the struggle for standards in this area is pointless if agreement on architecture and higher levels prevents shared use or interworking.

Consider for example the situation on ETHERNET where although both DEC and Xerox use the same media access protocols, their higher-level protocols are different. This means that such systems can share the same cable — a degree of integration — but without further agreement their office system products could not interwork. This leads to the situation where some LAN-based products should be considered to be 'closed' (ie they do not provide standard high-level protocols for interworking with other products).

LANs and OSI

Open systems can be achieved in two essentially different ways:

1 Agree on overall architecture and provide common higher levels (on top of each different basic LAN technology).

2 Hide the LAN technology behind converters which produce well known WAN-type interfaces.

Clearly, the first approach ought to be attractive since, in principle, the OSI concepts should apply equally well to LANs. In detail, the lower layers of LANs as defined by the IEEE 802 committees do not map cleanly onto the lower levels defined by OSI. However, there is an approach which provides unification above the network layer, ie at the OSI level 3 — level 4 interface (see Figure (4)14). The alternative approach of hiding the LAN is found in some products which offer 'interfaces' rather than applications.

An example is the Ungermann-Bass Net-One product *(024)*. Although the product uses the Ethernet standard (802.3) on the cable, the external user interfaces are normal V.24/RS232 type connectors (see Figure (4)15). The NIU boxes act as multiplexor devices, allowing interconnection of regular synchronous or asynchronous devices.

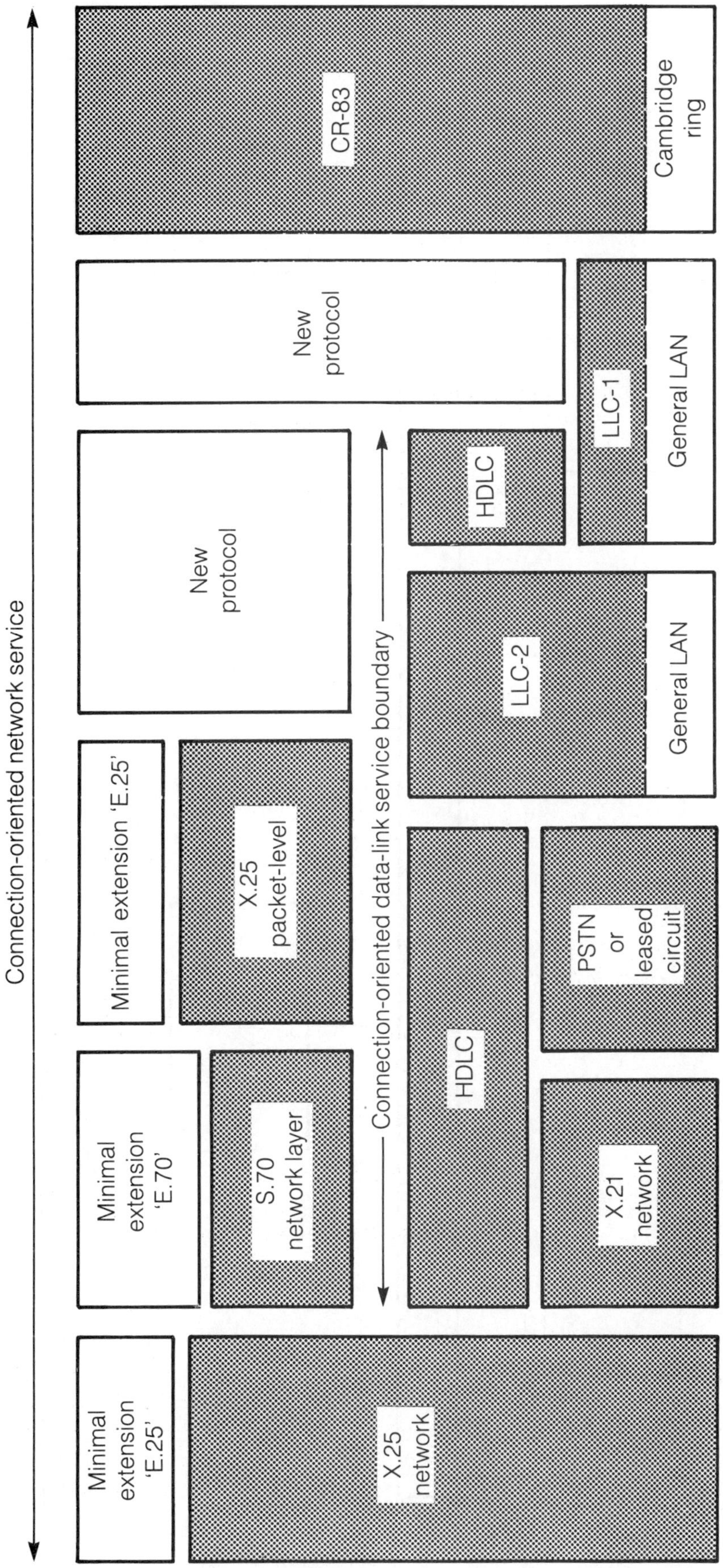

Note:
Shaded regions correspond to Intercept Recommendations; blank regions correspond to Interim Recommendations

Figure (4)14: Network protocol structure for the Intercept C-O Network Service

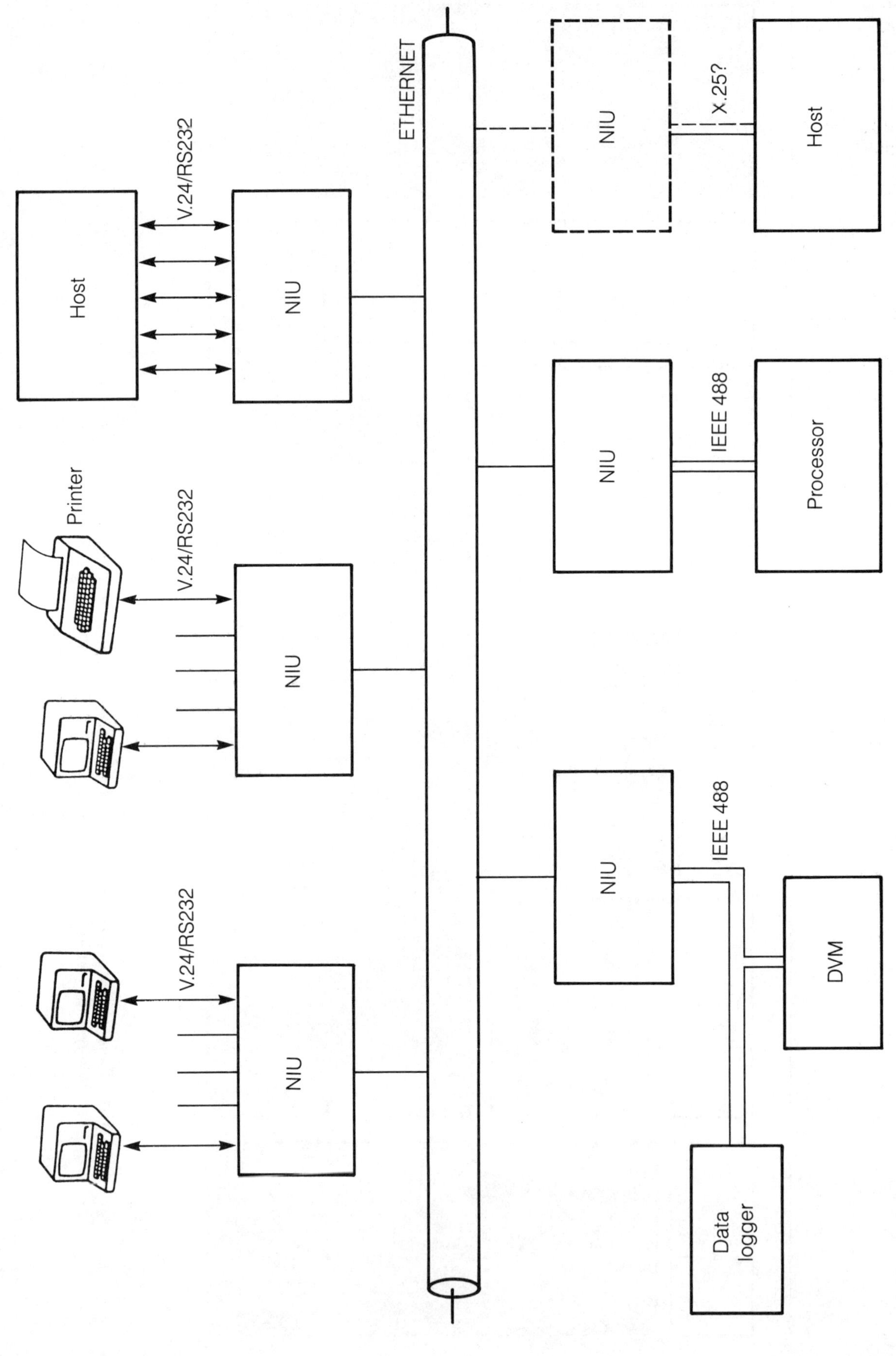

Figure (4)15: Invisible LAN

Other interfaces supported include the General Purpose Instrument Bus (IEEE 488) and it is understood an X.25 interface is being developed. The latter would allow the system to be used as a standard packet-switching network, offering WAN-type facilities but on a LAN cable.

LANs and gateways

In the same way as a PABX interworks with the public telephone network, so too, it must be expected that LANs will be required to interwork with WANs (and with other LANs across a WAN).

> *Stokes:*
> *Not only was it necessary to implement a LAN within St Thomas' Hospital but it was also necessary to provide links to other hospitals within the district, thus requiring interfaces to wide area networks.*

The two principal issues here are:

1 Architecture and common levels of protocol.
2 Network addressing.

As previously discussed, a gateway may be high level or low level. If the stations on the LAN use common (standard) protocols at level 4 and above, as shown in Figure (4)16, the gateway need only participate in the network layer conversions. However, if for some good reason (like high-performance requirements) the LAN stations use 'lean-mean' protocols, then it will probably be necessary to use a high-level gateway. It is unlikely that the high-speed close-coupled environment could be meaningfully extended over the WAN due to performance considerations. In this case it may be assumed that the bulk of the traffic is internal and that only a small proportion needs to traverse the gateway — which is just as well for performance considerations.

However, as Burrows states:

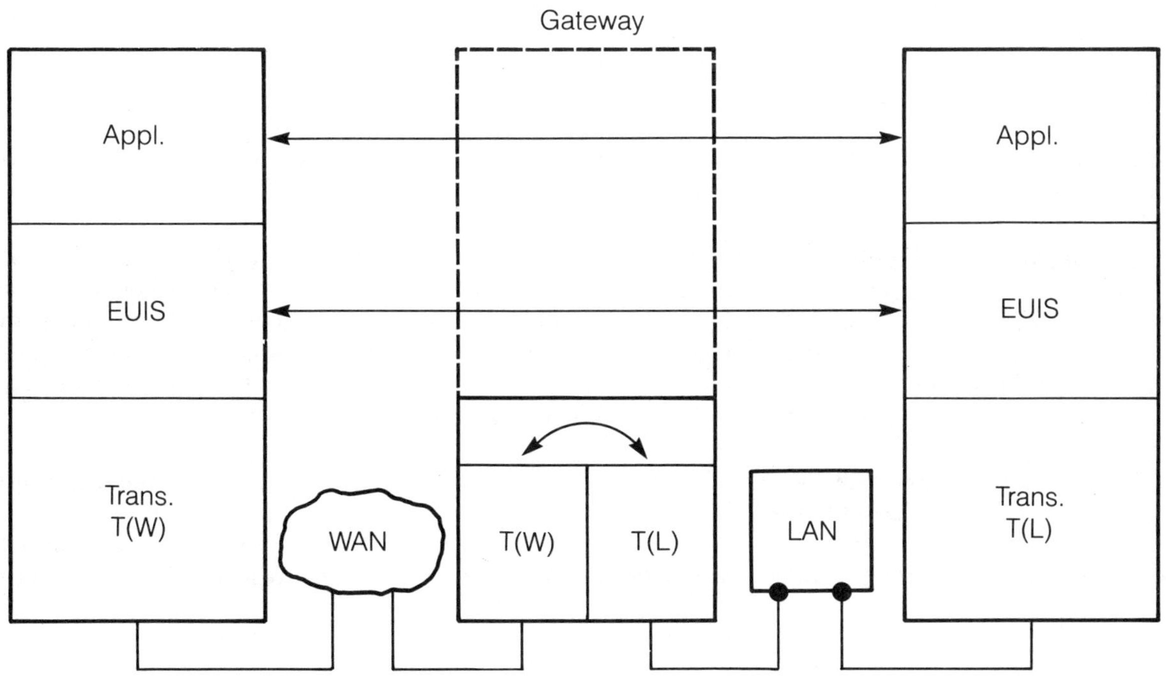

EUIS = End User Interworking Systems

Figure (4)16: Local-area/wide-area network gateway

Burrows:
The dividing line btween WANs and LANs is not obvious. For example, the plans to link universities into a communications network via the Project Universe scheme comprise a WAN; if a university used Ethernet to link its equipment together it would be using a LAN. It is likely that some central business districts will soon be wired up with fibre optics. Is this a LAN or a WAN? The author's view is that it is a LAN, but in order to encourage future development, a common terminology is needed. Many WANs using microwaves will be able to link into LANs using cable.

The second issue of network addressing again refers to the need for a universal numbering plan which allows subscribers on one network to access those on another. The issue is more emotive here since the WAN may be a public network whereas the LAN may be a private network.

On a PABX, a human operator is usually invoked to extend calls received from the PSTN (on the common number) to the required extension. Modern telephone networks permit Direct Dialling In (DDI) which allows the extension number to be dialled with the PSTN number. Alternatively an extension can be given a public network number of its own — this is called Direct Inward Dialling (DID). In either case operator intervention is not required.

This kind of operation will be essential in gateways between WANs and LANs for data calls; standards for general network services of this kind are proposed in ISO DIS 8208 for example.

Ettinger (2):
There may also be a need for connection to ISDN for another family of private multiservice equipment — LANs. However, their internal protocols are very specific and hence a gateway will be needed for their interconnection with the public network. Two solutions are at present envisaged:

1 By establishment of a transparent B channel in the public network and communication according to the internal protocol. This mode of operation, however, is possible only if both ends are compatible.

2 By conversion of the internal protocol to the standard protocol used at the public network interface. Such conversion can be simplified if compatibility is achieved at the LAN definition stage.

Other LAN technologies

Somewhere in the scale between the local area and the wide area lies the so-called Metropolitan Area Network.

The first appearances of this sort of network might be considered as the Community and Cable Television networks. These networks will enter many premises with very high bandwidth which, although primarily for 'broadcast' services, will also support two-way communication. The juxtaposition of these networks with other wide-area services will for some time only occur in certain areas, such as the larger conurbations. Specialist services (like those presently found in the City of London) can be expected to grow up in these areas but they may not be universally available for many years, if at all.

Other such technologies exist which should not be ignored, such as the plans for cellular radio and direct satellite broadcasting. The utilisation of these more esoteric media may be expected to be limited to specialist applications (eg police vehicles or ships at sea) at least for the next few years, until the technology and its harmonisation with the more common systems is readily available.

5: Strategy

In order to become the master of one's destiny, instead of being driven by events, it is necessary to devise a strategy concerning the processing and control of the organisation's information. Part of this strategy involves the integration of communication and computing services into a coherent whole. The first step is to realise, or convince those that need convincing, that a problem exists. This is the first of the seven pillars of wisdom, which are: why is the strategy necessary; whither lies the path; what aspects need tackling; when should things be done; who should carry out the strategy; which solutions are appropriate; how is the strategy to be organised; and how much will it cost?

Strategy

Getting started

Having recognised the problem, the next step is to formulate the strategy and this should be done in contemplative mood and with an enquiring frame of mind.

The questions to be answered are as follows:

- Why do we need to do something (status quo)?
- Whither lies the path (goals)?
- What aspects need tackling (scope)?
- When should it be done (phasing)?
- Who will take responsibility (staffing)?
- Which equipment is needed (technology)?
- How is it to be done (approach) and how much will it cost?

Why?

The answers to this question have already been hinted at in the preceding sections, but perhaps the most cogent reason is to be in control of events, rather than be controlled by them. A strategy will facilitate the anticipation of inevitable change so that future requirements can be more easily encompassed and not come as a surprise.

> *Dewis:*
> *Istel's information communication strategy was developed by its former Advanced Systems Department — a research and development function within the company with the brief to develop and maintain strategies in key areas, to evaluate new technologies and methods and back these with pilot projects where necessary. In asking the Advanced Systems Department to develop and implement such strategies, the company recognised the need and importance of these strategies and has already derived great benefits from their existence.*

If nothing else, a strategy should avoid the waste of following 'blind alleys'.

> *Stokes:*
> *There are two obvious ways of implementing such a model on a computer. First (classically), each separate activity may be a separate process on a large computer and these may be interconnected by interprocess calls and shared files. Secondly, each separate activity may take place on a different processor and these processors are linked by a communications network. It is the second of these approaches that was adopted by the strategic plan. There were various justifications for this approach. In particular, it allowed an incremental changeover, followed by incremental growth. Once the transition to a distributed network had occurred, there would be no need to replace a mainframe once every seven to 10 years — the various computers on the network could be replaced gradually, thus spreading out the capital costs.*

Whither?

In some sense, a strategy implies a goal. The goal may be to win the war, but to do this it is necessary to fight and win battles. Each battle will require its own (sub)strategy and the ultimate victory may not be clearly visible at the outset. However, it is appropriate to have some vision — a 'Holy Grail' — for which to strive (see Figure (5)1).

> *Dewis:*
> *A strategy can never be a statement of fact — it is a signpost pointing the direction towards a recommended destination which may be many years away.*

This may be seen as a unified system with a user enjoying:

- One access point having:
  - physical/electrical adaptors
  - link protocol adaptors
  - dynamic user-controlled sessions
  - multiple concurrent sessions
- One document structure containing:
  - file access adaptors
  - mixed mode documents (data, text, image, voice)
  - normalised formats
  - consistent document contents
- One command language with:
  - higher-level protocol adaptors
  - log-on password security
  - call progress signals
  - job control languages.

If this seems unattainable on a reasonable horizon then a likely path or trajectory needs to be established.

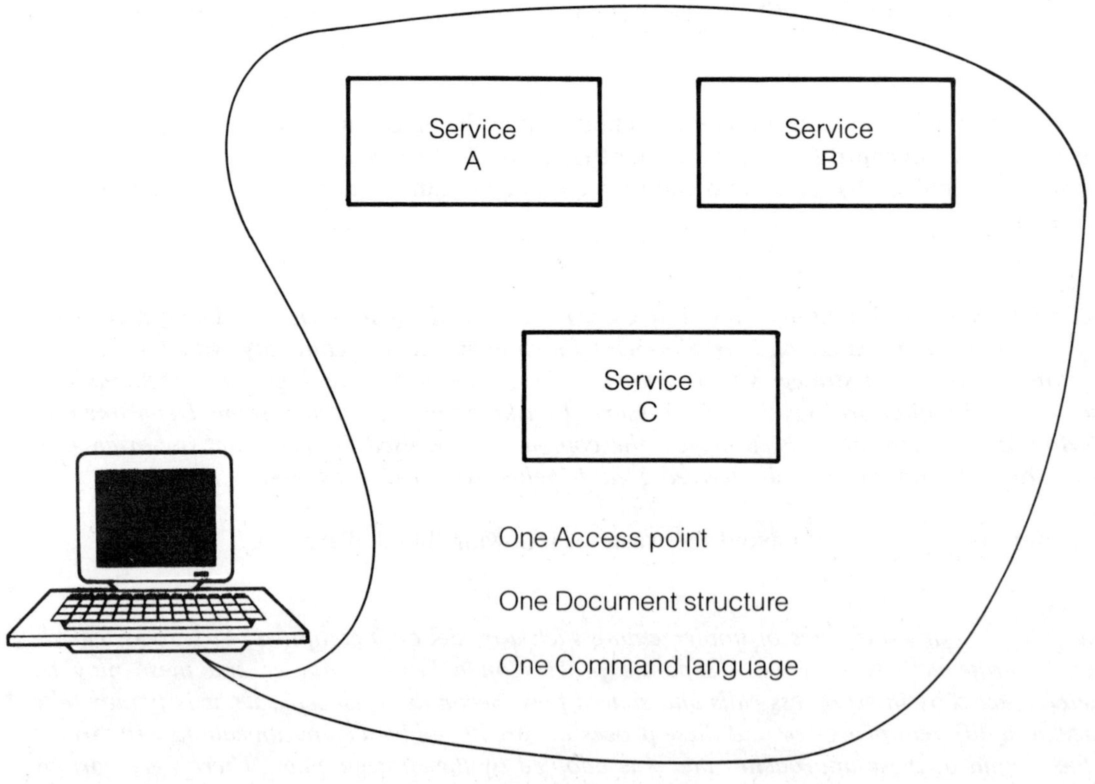

Figure (5)1: The 'Holy Grail'

Notice that, in this context, a user is anyone from the MD down who needs access to electronic information.

What?

The next question is: what facilities do the 'users' need? If this is less than the total set, then what are the necessary or possible subsets?

These questions must, of course, relate to the business requirements of the organisation.

> *Dewis:*
> *It is important that every company, whatever its size, should develop IT strategies to address its business needs and thus be in a position to make maximum use of technological developments as they become appropriate.*

In order to establish the operational requirements of future systems it is necessary to:

1 Find out:
- Key functions
- Locations
- Equipment installed.

} Now, planned, tentative

2 Assess:
- Communications mode(s)
- Traffic volumes
- Performance levels.

} Now (measure) Future (estimates)

3 Evaluate:
- Technology constraints
- Problem areas
- Full training needs.

The functions involved may be many and various including not only person-to-person but also the use of computer-based functions such as:
- Expert systems
- Database and information systems
- Production control
- Process control
- Telemetry and telecontrol
- Financial systems (cash management, Electronic Funds Transfer (EFT) etc)
- Calculation applications (payroll, stress analysis, CAD/CAM)
- Office systems (WP, filing, managing, personal computing)
- Systems development etc.

When?

In addition to establishing the user needs, it is necessary for reasons of practicality to establish by when certain needs must be met and, if applicable, any priorities. 'Needs', of course, must not be confused with 'wants' and financing may restrict the introduction of particular facilities. It also happens that individuals do not know what they want until they see it!

In planning for the future it must be recognised that the business cannot stand still until the strategy is in place. This means that some short-term steps must be determined while the longer term issues are being identified.

There may be other time constraints such as rate of investment in equipment or availability of technology. These will lead to the need for a phased approach going by a series of steps towards the next goal. However, it is only possible to plan so far ahead because of business and technology horizons (typically, say, five years). This means that the strategy must continually be under review (see Figure (5)2).

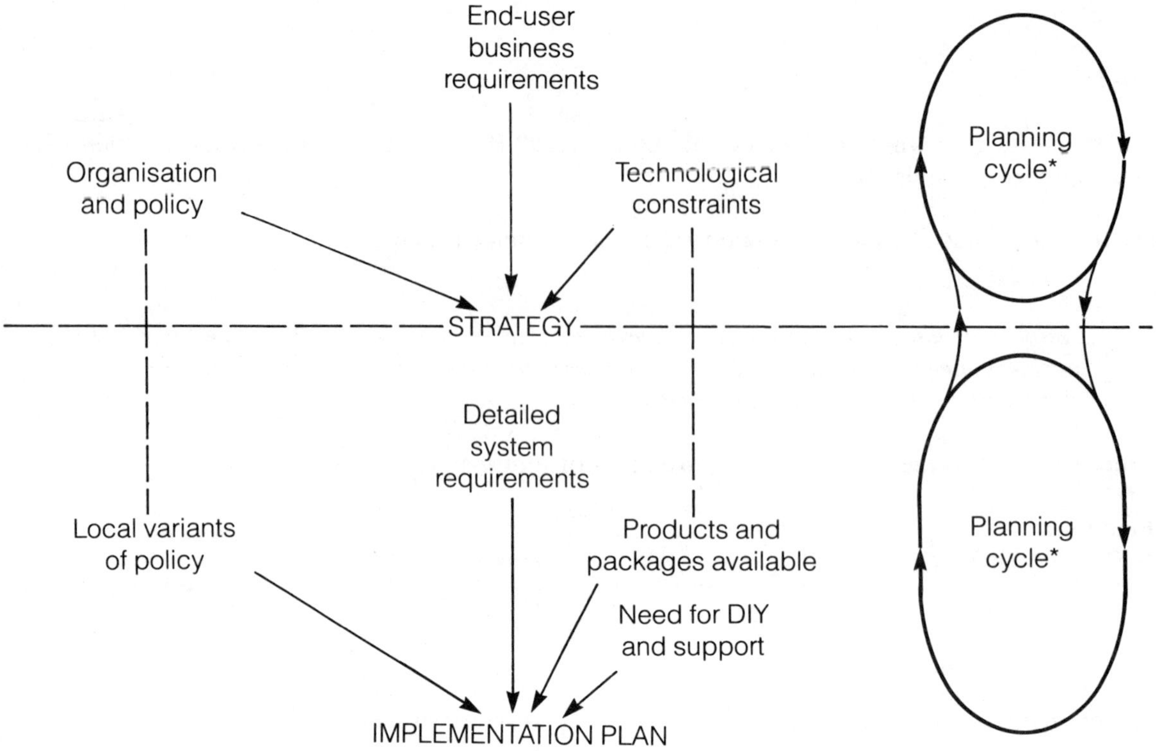

**Planning really means planning and monitoring*

Figure (5)2: Strategy plan

> *Dewis:*
> *A strategy may never be accomplished totally. Views may be modified based on two factors — changing technology resulting in new or more cost-effective facilities and changing end-user requirements which can be affected by user awareness of systems capabilities. Either of these could render parts of the strategy inapplicable. Consequently, Istel believes that its key strategies will need to be reviewed and perhaps refined every two or three years.*

Who?

It is obvious that formulation of a strategy along the lines indicated here must be instigated at a high level in the organisation. It must be part of corporate policy to have any hope of success. The corporate policy is determined by business considerations and this frames the requirements that must be met at the functional level (see Figure (5)3). This implies an identified organisational responsibility and structure. If these functions are presently dispersed (see Figure (5)4), then the organisation must be put in place (see Figure (5)5).

The staffing for this function may be drawn from internal resources, or may have to be recruited (particularly if the expertise is not in-house). Alternatively, the organisation could put the whole matter in the hands of a manufacturer, although this may seem a little risky.

Since getting ahead of the game implies an extra effort for a limited period, the necessary, temporary increase in staffing levels can be met, ideally perhaps, by resorting to expert help in the form of consultant or contracted staff.

Which?

The equipment issue is always emotive. Fierce brand loyalties are formed within different parts of an organisation, existing investments or contracts reduce the options available and everybody is an expert in their particular area. The technological scene is also changing at an alarming rate and is viewed both as a threat and as an opportunity.

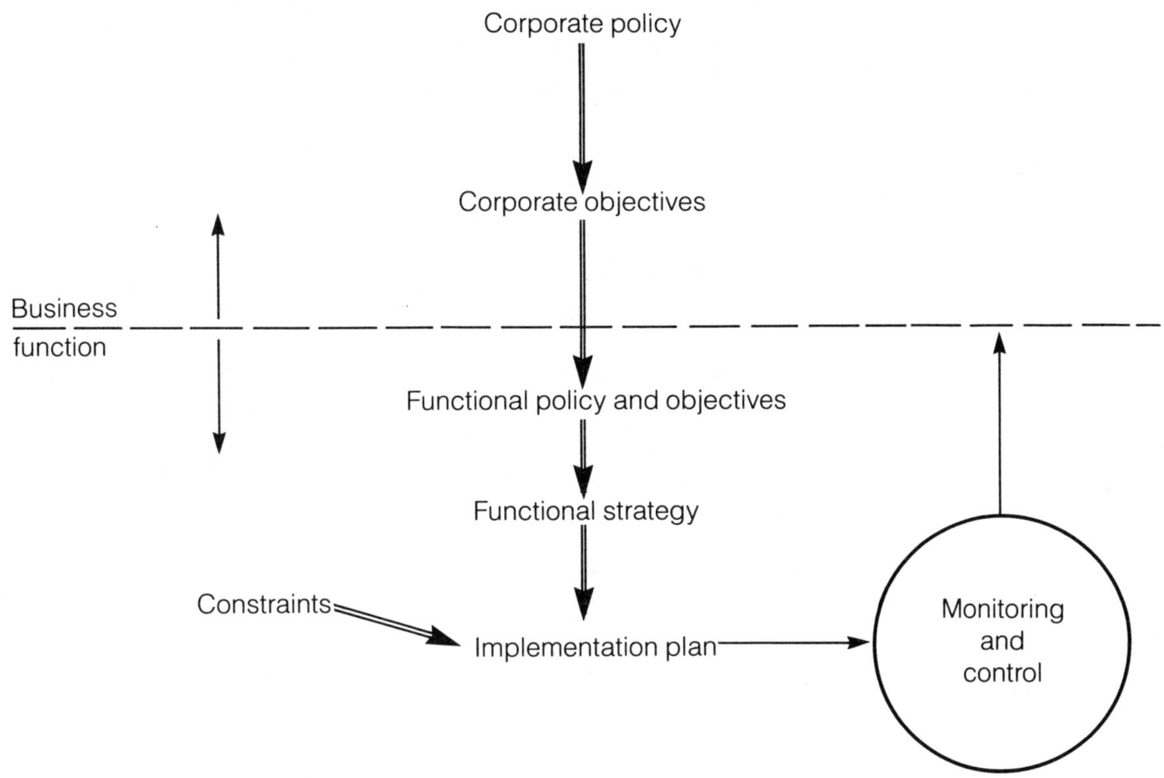

Figure (5)3: Factors affecting implementation

Communications service	Controlled by
DP	1 DP manager 2 Microsystem users
WP	1 Office manager or DP manager 2 Microsystem users
Telex	1 Office manager or site communications manager
Voice	1 Site communications manager or office manager
• Fragmented • No integration • Firefighting	

Figure (5)4: The organisation now

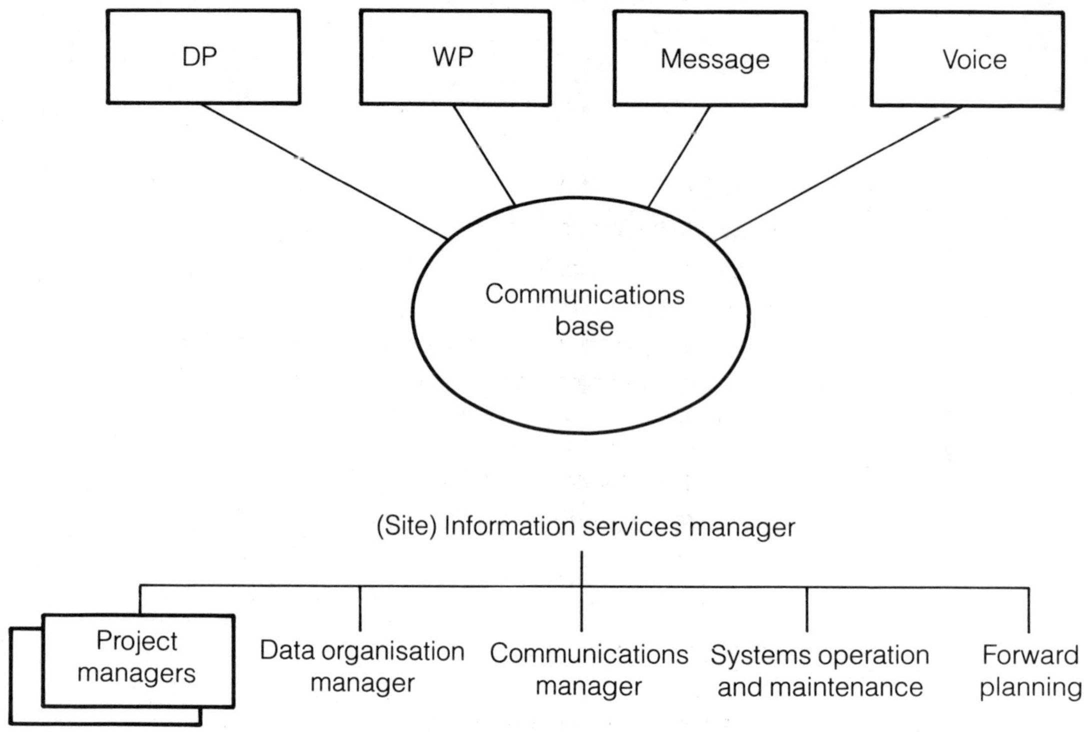

Figure (5)5: The organisation soon

This is the area where mistakes are easily made since products are seen as short-range solutions to tactical problems. It is also the area where manufacturers do battle with each other (and the customer).

Equipment purchase must take second place to strategy — not lead it. Strategy development should:

1 Firm up on communication mode(s), traffic volumes and projections, security/privacy requirements, performance levels, by consultation with those responsible.

2 Agree policy on technology:
- Trail blazing or conservative (let someone else find the bugs)
- Single or multiple supplier
- Open or closed systems.

3 Assess constraints:
- Current applications
- Expertise available
- Geography
- Money.

Central to this, and the one thing that must evolve ahead of demand, is the telecommunication strategy (see Figure (5)6).

How?

How is all this to be done? The practicalities do depend heavily on the organisation and the nature of its business. As will all these things, the key often lies in determining the user requirements which should lead to the underlying information flow requirements. From this, the traffic flow matrices can be educed which become the input to the communications strategy. All that is needed then is an implementation plan and the organisation to keep it running smoothly. These tasks are not simple,

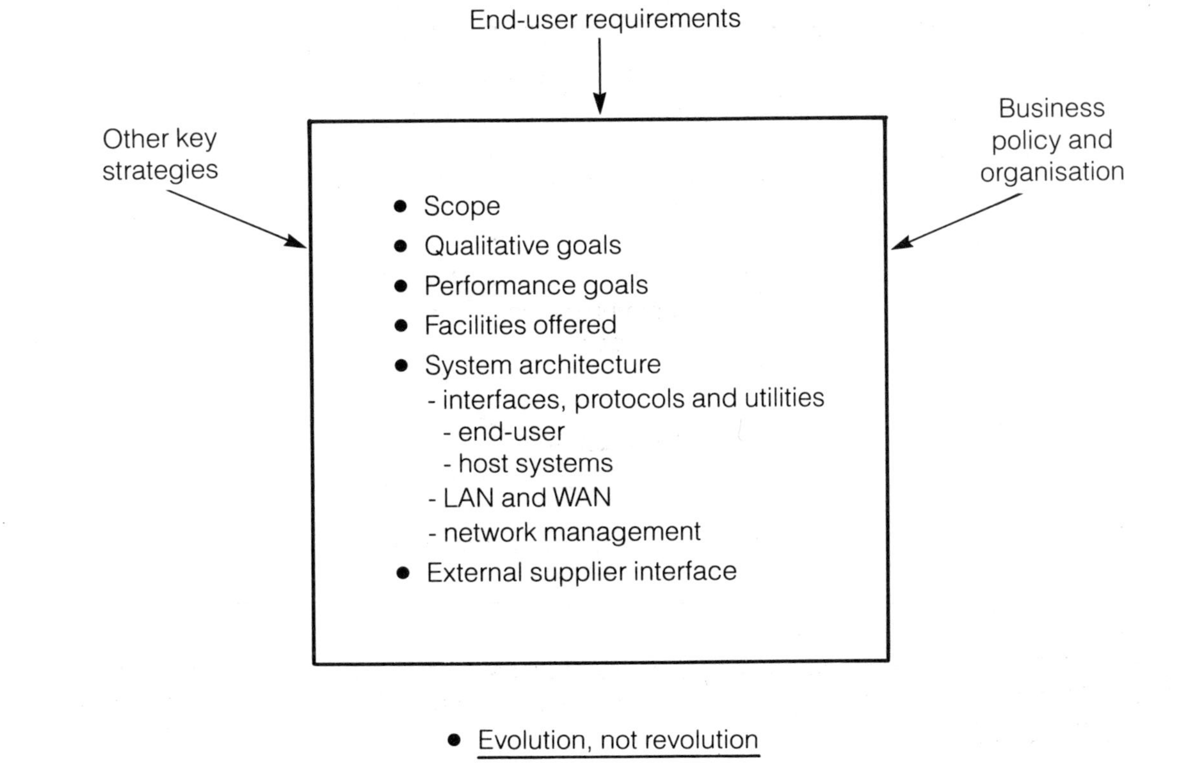

Figure (5)6: Communications strategy definition

even for those experts who are versed in the arts, but a number of tools are used such as:

- Semi-structured interviews
- Interworking requirements matrix
- Generation and comparison of scenarios
- Kepner-Tregoe *(025)*.

Implementation options include:

- The classical 'big bang' (appropriate to situations where implementation can start from the beginning)
- Prototypes (limited to certain functions at first)
- Pilot systems (limited to certain sites at first)
- Experiments (limited set of comparisons).

The areas of cost can be broadly identified as follows:

- Establishing user requirements
- Determining priorities
- Producing functional specifications
- Issuing invitations to tender
- Evaluation of alternatives
- Installing new systems
- Operations, including maintenance, monitoring and training
- Planning extensions, expansions and replacement.

Exactly how much the total cost will be is entirely dependent upon *how much you wish to spend*.

Analysis references

001
Meade D (editor)
'Microprocessor systems'
State of the Art rep series 11 no 2 p 190
Pergamon Infotech (1983)

002
Birrell A, Levin R, Needham R and Schroeder M
'Grapevine: an exercise in distributed computing'
Communications of the ACM vol 25 no 4
(Apr 1982)

003
'Mailway: capabilities guide'
Wang document 700-5979A

004
Wilson P
'Let's get mailboxes moving'
Computer Weekly p 26 (1 Mar 1984)

005
Clark R T
'CHAPS: a new approach to payment systems'
Inter-Bank Research Organisation (1982)

006
International Data Corporation
Survey in EDP Industry Rep (21 June 1982)

007
'ISDN — a description'
Issue 3 Network Executive
Network Strategy Dept British Telecom
(Aug 1982)

008
Dewis I
'The British Steel network'
NATO Advanced Study Institute: on interlinking of computer networks
series C no 12 pp 345-352 D Reidal Publishing Co
(Aug 1978)

009
'New horizons in multiplexing'
Communicate p 28 (Nov 1983)

010
Kingsmill A
'A PABX-based integrated business information system'
Proc LocalNet '83 (Europe) Online Publications Ltd (1983)

011
IBM documents:
GC30-3085 version 3 (DISOS, System 370)
GC23-0765 (Offices information architectures: concepts)
SC23-0759 (DIA)
SC23-0757 to 0758 (DCA)

012
'Information processing systems — Open Systems Interconnection — basic reference model'
International Organisation for Standardisation
IS7498

013
Meijer A and Peeters P
'Proprietary network architectures and OSI'
In 'Network architectures' C Solomonides (ed)
State of the Art rep series 10 no 1 Pergamon Infotech (1982)

014
Gurugé A
'SNA: theory and practice'
Pergamon Infotech (1984)

015
'X.25: an introduction to the X.25 recommendation'
Logica Ltd 64 Newman Street London W1A 4SE

016
Post Office Electrical Engineers J
(collection of papers on System X) British Telecom (1979-1982)

017Dewis I
'Teletex — setting the standard for electronic mail?'
In 'Network architectures' C Solomonides (ed) State of the Art rep series 10 no 1 Pergamon Infotech (1982)

018
Heys F
'Videotex networks: usage, approaches and interconnection'
In 'Network architectures' C Solomonides (ed) State of the Art rep series 10 no 1 Pergamon Infotech (1982)

019
Bennett P
'London stock exchange bullish about videotex'
Data Communications McGraw-Hill (Aug 1981)

020
'Document interchange protocol for the telematic services'
CCITT recommendation T73 ITU Geneva (Apr 1984)

021
Rance J
'IEEE project 802 — a status review'
Proc LocalNet '83 (Europe) p 399 Online Publications Ltd (1983)

022
'Cambridge Ring '82 protocol specifications'
Joint Network Team of the Computer Board for Universities and Research Councils (1982)

023
Lees W
'The Universe network at Logica'
Proc LocalNet '83 (Europe) p 311
Online Publications Ltd (1983)

024
Marshall J and Spiegelhalter B
'Experiences with Net-One at British Telecom'
Proc LocalNet '83 (Europe) p 67
Online Publications Ltd (1983)

025
Kepner C and Tregoe B
'The rational manager, a systematic approach to problem solving and decision making'
McGraw-Hill (1965)

026
'How to succeed in the X-Stream age'
Timeplex Business Communications
Timeplex Ltd 77 Boston Manor Road
Brentford TW8 9IQ

Bibliography

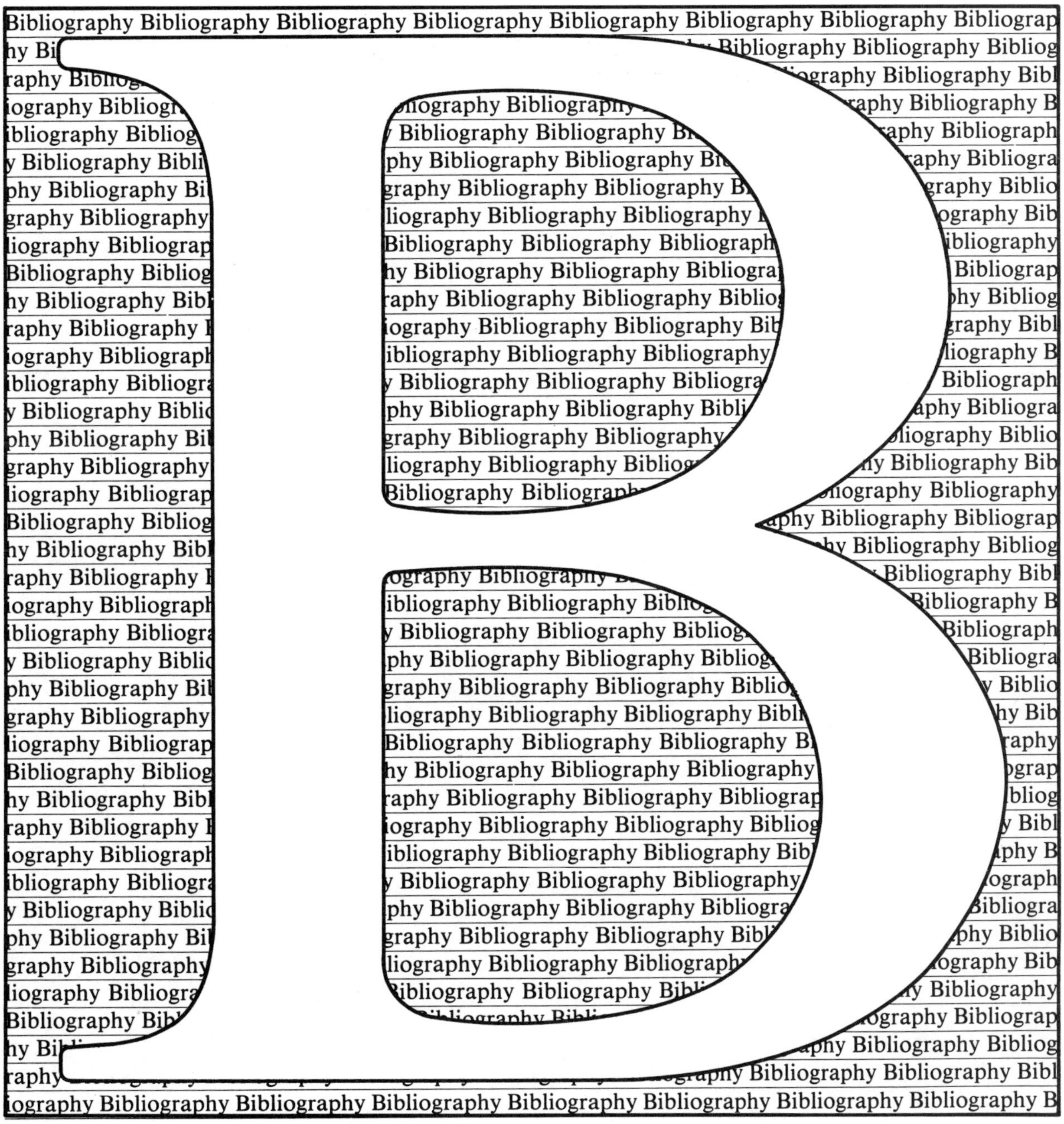

An annotated bibliography of strategies for integrated communications

B C Burrows

Futures Information Associates
Milton Keynes
UK

An annotated bibliography of strategies for integrated communications

Introduction

There is a need for more publications which show the converging relationship between new forms of communications, computers, information storage, retrieval and transmission and the impact of this on the structure of commerce and society and changing work patterns.

There is a wide range of publications for the telecommunication engineer from elementary to advanced level. Until recently there has been little relationship between this group and other professions except for people engaged on research in information theory. This, which some have argued should be called communication theory, is the mathematical analysis of information flow in communication channels by means of electrical transmission. Attempts have been made to relate this to the exchange of information between people and organisations with little success. We need more research on the interaction between closed and open systems.

This bibliography has been arranged alphabetically by the author. It was decided not to group publications by subjects because all the items selected are either closely related to other subjects or could soon be related. It is likely that with converging technology we will all have a black box which will combine the functions of a television set, a computer, a telephone and an interactive terminal as well as being a means of teleshopping and banking. As the situation is fluid the basis of selection has been an attempt to reflect all these trends.

The basis of selection has been as follows:

1 The traditional communications literature.
2 Computers and networking.
3 Protocols and interfacing.
4 Cable development and the concept of the wired city and nation.
5 Microwaves and satellites.
6 Development in the use of fibre optics.
7 Studies of convergence of systems.
8 Future projections.
9 Social and business implications of the new technology.

All these articles have been selected from books, published conference proceedings and research reports. There are also a large number of periodicals covering this subject — therefore a small representative selection is made from these so that this Report can be used as a basis for keeping up to date on this complex and rapidly changing subject. It is vital that we do so because the business opportunities are very great for whoever wins the race for the most advanced integrated communications system.

Abstracts

100
Aldrich M
'Videotex — key to the wired city'
Quiller Press 115 pp (1982)

The author who is chief executive of Rediffusion Computers looks at the developments in videotex and sees this as the key to the development of cable, new forms of entertainment, the electronic office and the wired city and the wired nation.

101
Cawkell A E
'An investigation of commercially available facsimile systems'
British Library Lending Division R&D Rep 5719 (1983)

This report gives information about the performance and price of commercially available facsimile systems and supplements this data with both technical and social factors. A review of the market and the potential of facsimile in relation to new development is made. A glossary of terms is included.

102
Centro Studi E Laboratori
'Optical fibre communications'
McGraw-Hill 928 pp (1984)

CSELT is the central research laboratory for some 20 Italian companies in the telecommunications sector. This is a reference book on design, production, installation, operation and use of optical fibre systems.

103
Chou W (editor)
'Computer communications vol 1 Principles'
Prentice-Hall Inc (1983)

This work examines the development of computer communications networks. The book is a series of chapters from experts in their field on radio links, satellite channels, coaxial cable and fibre optics. ARPANET and IBM's SNA are described as well as network security.

104
Chou W (editor)
'Computer communications vol 2 Systems'
Prentice-Hall Inc 496 pp (1984)

This volume contains a series of papers which describe the latest developments on network, data/voice/image systems, distributed database design, digital developments and package and circuit switching.

105
Communications and Information Technology (CIT) Research Ltd
'Cable TV communications in Western Europe'
Communications and Information Technology (CIT) Research Ltd 20 pp (1982)

This is one of the first reports to analyse and describe the concept and problems of cable TV in Europe. The conclusions are that CATV is speculative and its economies not easy to assess. However, restrictions should be lifted so that the service can develop.

106
Corby M and Donohue E J
'The telecomms user's handbook'

Telecommunications Press 320 pp (1983)

This work, written by the Directors of the Telecommunications Users Association, provides a practical guide to telecommunications equipment, services and terminology. The book provides useful information for managers, administrators and executives as an aid for installing telecommunication equipment.

107
Cornish E
'Communications tomorrow: the coming of the information society'
World Future Society 160 pp (1982)

An anthology of articles from the Futurists covering such topics as computer conferencing, home computer terminals, new methods of communication and the development of the information society.

108
Cypser R J
'Communications architecture for distributed systems'
Addison-Wesley Publishers Ltd 711 pp (1978)

Mr Cypser deals with the area of communications among distributed systems. IBM's System Network Architecture (SNA) is described in detail. Interfacing to other networks is described and an overview of operations is given.

109
Davies D W, Barber D L A, Price W L and Solomonides C M
'Computer networks and their protocols'
John Wiley and Sons Ltd 504 pp (1979)

The development of networks has focused attention on formal protocols and standard interfaces and these are fully described in this book.

110
Davies D W and Barber D L A
'Computer networks for computers'
John Wiley and Sons Ltd 575 pp (1973)

This work gives a comprehensive overview of computer networks and the technology created by the convergence of computing and telecommunications.

111
Davies E
'Telecommunications — a technology for change'
HMSO 60 pp (1983)

This booklet was published to coincide with the opening of a gallery on telecommunications in the Science Museum. It traces the history of telecommunications and how these have converged into the technology used today. Although brief, it is well illustrated and forms a useful introduction to the subject.

112
Deasington R
'A practical guide to computer communications and networking'
John Wiley and Sons Ltd 132 pp (1982)

This work attempts to bridge the gap between the electronic engineer and the computer scientist. It deals with the physical mechanisms of data transmission and the writing of communications software.

113
Deasington R
'A practical guide to computer communications and networking'
Ellis Horwood Ltd 132 pp (1982)

This book is a practical guide on the use of communications software in order for computers to communicate with other computers or terminals.

114
Defler F and Skillings W
'A managers' guide to local area networks'
Prentice-Hall Inc 253 pp (1983)

This work deals with local area networks from the users' point of view. Although some of the language is technical it is always preceded by a description of what the LAN or its components are. It is, therefore, a useful introduction.

115
Department of the Environment and Transport
'Inputs of telecommunications on planning and transport'
Department of the Environment and Transport Research Rep 272 pp (1978)

A series of very useful papers concerned with the impact of telecommunications on business and government, transport, office location and welfare in the community.

116
Didsbury H F (editor)
'Communications and the future: prospect, promises and problems'
World Future Society 357 pp (1982)

A selection of papers covering a wide range of

topics. Most of the papers were presented at the World Future Societies General Assembly in Washington DC in July 1982.

117
Economic Intelligence Unit
'Cable television in Western Europe — a licence to print money'
Economic Intelligence Unit (1983)

The report argues that the development of cable television will be both fast and profitable due to the demand for multichannel television.

118
Economic Intelligence Unit
'The role of communications in human settlement development'
(Sponsored by the United Nations Centre for Human Settlement (HABITAT))
Economic Intelligence Unit 39 pp (1983)

Human settlement is described in relation to telecommunications, telecommunication economics and policies for human settlement development.

119
Edgar P and Rahim S A (editors)
'Communication policy in developed countries'
Kegan Paul International 297 pp (1982)

The technological and cultural forces shaping policy are examined in this book. This is reviewed in relation to international aspects and an attempt is made to examine a conceptual framework as an aid to future developments.

120
Ekistics
'Communications with and without technology'
Ekistics vol 50 no 302 pp 317-421 (Sep/Oct 1982)

This is a collection of articles attempting to explore the nature of communications between people. Both the social aspect of communications networks and their development via telecommunications are described.

121
Elton M C J
'Teleconferencing: new media for business meetings'
American Management Associations 57 pp (1982)

This booklet consists of a brief introductory chapter, a long chapter on audio conferencing and further sections on video conferencing and the relationship between teleconferencing and travel. Its aim is to explain the use and development of teleconferencing for the manager.

122
'Television: the new era'
Euromonitor Publications (1983)

This report argues that because of the wide use of video recorders in the UK, which is the highest per head of population in relation to the rest of the world, that demand for multichannel TV via cable and satellite will not be as great as estimated.

123
Feher K
'Digital communications: satellite/earthstation'
Prentice-Hall Inc 496 pp
(1983)

This work gives a comprehensive coverage of modern digital communication systems. The descriptions of satellites and earthstations are international in scope and deal with both operational and planned systems.

124
Feher K
'Digital communications: microwave applications'
Prentice-Hall Inc 269 pp (1981)

The concept and technique of digital systems by line-on-sight microwaves is described here. Both industrial and academic achievements are dealt with. Numerous practical examples are given ranging from digital modulation schemes to filtering techniques.

125
Flint D C
'The data ring main: an introduction to local area networks'
John Wiley and Sons Ltd 375 pp
(1983)

The theory, technology, protocols and interfacing techniques are described. The implications of their use are discussed and details of available LANs are given.

126
Future Systems Inc
'Optical fibre communications'
Future Systems Inc Rep no 116 135 pp
(1982)

The report covers a broad spectrum of fibre optics from laser concept to cost tradeoff between microwave and satellites. It is estimated that the market will grow to about $1 billion by 1985, $3 billion by 1990, $5.7 billion by 1995 and $6.3 billion in the year 2000.

127
Gassmann H-P
'Information, computer and communications policy for the Eighties'
North-Holland Publishing Co 276 pp (1983)

An OECD report which reviews work done in Europe and suggests guidelines for future development.

128
Gecsei J
'Architecture of videotex systems'
Prentice-Hall Inc 336 pp (1983)

Part I of this book is an introduction to videotex and computer networks, Parts II to IV deal with the architectural layers of the ISO Open Systems Interconnection model. In the final section on applications database, teletext and gateway are described as well as future developments in alternative methods of interfacing and telesoftware.

129
Gee K C E
'Introduction to local area networks'
Macmillan Press (1983)

The concept of the local area network is described including the differing types such as star, bus and ring. The main sector deals with both hardware and software requirements as well as performance and applications of LANs.

130
Gee K C E
'Local area networks'
National Computing Centre 253 pp (1982)

This book describes local area networks and their application in detail. Recent developments, the use of PABX, the problem of standards and the choice of network as well as future developments are dealt with. This publication is useful for designers of computer and office systems.

131
Goddard J B and Morris D
'The communications factor in office decentralisation'
In 'Progress in planning' vol 6 part 1 80 pp Pergamon Press (1976)

This work deals with the role played by communications when offices are moved from urban areas such as London. It forms a useful background for planning a telecommunications service.

132
Golzen G (editor)
'The new technology at work'

The Telecommunications Press 200 pp (1982)
This is the printed version of a symposium held at the Barbican Centre in 1981. It covers the broad issues of how the new technology will change the way people will work and the impact it will have on the design of office buildings.

133
Graham J
'The Penguin dictionary of telecommunications'
Penguin Books 199 pp (1983)

This guide to terms used in communications is both comprehensive and up to date. As well as 1500 defined terms relating to telephone, radio, data transmission and television systems, there is a useful technical series of appendices.

134
Grange J L (editor)
'Satellite and computer communications'
North-Holland Publishing Co 380 pp (1983)

This publication contains the papers of an international conference sponsored by IFIP Technical Committee No 6. There are 27 papers by experts from more than 10 countries.

135
Glossbrenner A
'The complete handbook of personal computer communications: everything you need to go on-line with the world'
St Martins Press 325 pp (1983)

This is the first comprehensive work which attempts to link the use of the personal computer to the vast range of on-line information. It is written from the point of view of the personal computer owner but reflects US practice.

136
Greater London Council
'Cabling in London'
Rep by Economic Policy Group Greater Urban London Rep no IEC 586A 125 pp (1982)

This report deals with the history of cable, its uses both for commercial and public service information and its relationship to the development of cable TV including public access and education. An appendix deals with the Hunt report, cabling in Britain and in other countries. This is one of the most detailed reports on cable.

137
Haugh R W, Gerbner G and Byrne R B
'Communications in the twenty-first century'
John Wiley and Sons Ltd 240 pp (1981)

An attempt to assess the effects of electronic communication systems on society in the future is made in this book. Examples are given of the advantages available in the use of improved communication systems.

138
Griffiths J M (editor)
'Local telecommunications'
Peter Pereginus 265 pp (1983)

This work examines the present state of the local area network and considers how it will evolve and integrate with other services.

139
Haykin S
'Communication systems'
John Wiley and Sons Ltd 653 pp (1983)

This work examines in practical detail the principles of communications theory as applied to the transmission of information with equal emphasis on analogue and digital systems.

140
Hiltz S R and Turoff M
'The network nation: human communications via computer'
Addison-Wesley Publishers Ltd 528 pp (1978)

This book gives an overview of computerised conferencing, including its social and psychological aspects, opportunities and future development. Although one of the first works on this subject it is still useful.

141
Home Office and Department of Industry
'The development of cable systems and services'
HMSO 89 pp (1983)

This is the report recommending the future of cable development within the UK. Based on the ITAP and Hunt report the conclusion is that cable TV will be led by the leisure sector before the business sector. There has been a series of reports since this publication both supporting and disputing this view.

142
Hopper A and Wheeler D J
'Local area networks: the Cambridge Ring'
Addison-Wesley Publishers Ltd 256 pp (1983)

The Cambridge Ring was developed at Cambridge University and has become the basis of Project Universe. This work by two members of the Cambridge Ring team describes the setting up, design and operation of this network.

143
Housley T
'Data communications and teleprocessing systems'
Prentice-Hall Inc 262 pp (1979)

This work deals with the fundamentals of data communication, details the reasons for errors in the system and explains the terminology. Planning a system, network protocols and common and problem areas are also covered.

144
D Bates (editor)
'Network systems and software'
Infotech State of the Art Rep Pergamon Infotech 692 pp (1975)

Network systems, their development and their relationship to software are dealt with in this report.

145
Solomonides C (editor)
'Network architecture'
Infotech State of the Art Rep series 10 no 1 379 pp Pergamon Infotech (1982)

This report deals with ISO layers, network architecture, Teletex and electronic mail, videotex and network protocols. All aspects are examined in relationship to the ISO layers.

146
Stokes A V (editor)
'Networks'
Infotech State of the Art Rep series 9 no 2 422 pp Pergamon Infotech (1981)

The network planning, architecture and standards are described here. Papers are included on European telecommunications, networks in the 1980s and local area networks. Future development and plans are described.

147
Johanson R, Vallee J and Spangier K
'Electronic meetings: technical alternatives and social choices'
Addison-Wesley Publishers Ltd 244 pp (1979)

This work deals with the impact of new forms of electronic methods such as audio, video, computer or teleconferencing from the point of view of the meetings themselves. These techniques are explored as alternatives to face-to-face meetings. A useful work in an area of change and uncertainty.

148
Kerr E B and Hiltz S R
'Computer-mediated communication systems: status and evaluation'
Academic Press (1983)

This work attempts to synthesise the current state of research in the areas of computer conferencing, electronic mail, automated office and general communications systems.

149
Lancaster K L (editor)
'International telecommunications: user requirements and supplier strategies'
Lexington Books 197 pp
(1982)

This book contains 12 papers all by employees of Arthur D Little Inc. It provides an overview of information useful to large suppliers of telecommunication equipment in the European market.

150
Lane J and Scott P
'Guide to British Telecom Services'
National Computing Centre 70 pp
(1983)

A guide to the services from British Telecom particularly in the field of data transmission. Future and planned services are described.

151
Lee A M
'Electronic message transfer and its implications'
Lexington Books 195 pp
(1983)

All aspects of electronic message transfer including its implication for labour and employment, privacy and future development are dealt with in this publication.

152
Lehman M and Burke T J M (editors)
'Communication technologies and information flow'
Pergamon Press 168 pp (1981)

A series of papers dealing with the development of most aspects of the new communications technology including the electronic office, satellite communications, national and international developments as well as domestic use.

153
Leinwoll S
'From spark to satellite — a history of radio communications'
Charles Scriber's Sons 242 pp (1979)

A readable history of the development of radio and its recent and future application. The use of lasers is seen as playing a major role in the future.

154
Logica Ltd
'Communications in Europe — the changing environment'
Logica Ltd 300 pp (1983)

The view is expressed that the current uncertain investment pattern will be replaced by much steadier and stronger growth during the next decade. This conclusion is made after a study of the plans of 14 European countries. In the final section developments in satellite communications are described and the regulations in relation to telecommunications are examined.

155
Logica Ltd
'The data communications market in Western Europe 1981-1987'
Online Publications 179 pp (1981)

This study was conducted by Logica over a period of two years and the results given at an on-line conference in 1981. It is estimated that the total number of terminals will rise by more than six times the 1979 level by 1987 and 67 per cent of these terminals will be used for data communications.

156
Look H E (editor)
'An introduction to electronic publishing: a snapshot of the early 1980s'
Learned Information 200 pp
(1983)

There is a large amount of literature in electronic publishing. The aim of this work is to describe the industry in the 1980s which includes on-line, videotex and the telecommunications industry. The relationship of electronic publishing and expert systems and local area networks is mentioned. A useful introduction to the literature on this subject and how it relates to new forms of communications is included.

157
Loomis M E S
'Data communications'
Prentice-Hall Inc 256 pp (1983)

This book deals with the history of data communications, use of terminals, networking, data codes and protocols and software developments. Although this work is aimed at the student the information given is useful in order to help establish an overview of the new developments taking place.

158
Martin J
'Communications satellite systems'
Prentice-Hall Inc 398 pp (1978)

This work explores the field of electronic communications and how this will develop via the use of satellites. Like most of this author's work this is an interesting introduction to the subject.

159
Martin J
'Computer networks and distributed processing'
Prentice-Hall Inc 562 pp (1981)

The potentials of network systems are described as well as a general description of network mechanisms and the standards required for interfacing. This work concludes with the problems of error control and security.

160
Martin J
'Design and strategy for distributed data processing'
Prentice-Hall Inc 624 pp (1981)

The design, potential applications and method of distributed data processing systems are described. The concluding section deals with software, network strategy and security systems.

161
Martin J
'Future developments in telecommunications'
Prentice-Hall Inc 668 pp (1977)

This book deals with the communications revolution and future developments. It covers all aspects of communications from telephone, radio networking, cable, satellites, package switching and the concept of the wired city and working from home. A very useful introduction to this wide-ranging subject.

162
Martin J
'Introduction to teleprocessing'
Prentice-Hall Inc 267 pp (1972)

A useful introduction to data transmission techniques from their beginning. Equipment needed is described as well as software structure, flow of data and design techniques. An appendix gives comprehensive summaries and a checklist for use in selection of equipment. A new edition would be welcome but this is still a useful work.

163
Martin J
'Systems analysis for data transmission'
Prentice-Hall Inc 909 pp (1972)

This work concentrates on the relationship between software and system design and user considerations and the equipment available. It covers this subject in some detail.

164
Martin J
'Telecommunications and the computer'
Prentice-Hall Inc 670 pp (1976)

This is the second edition of a successful work on this subject. It deals with man-machine dialogue, the structure of telecommunications networks, pulse code modulation, switching techniques and the problems of noise and distortion.

165
Martin J
'Telematic society. A challenge for tomorrow'
Prentice-Hall Inc 244 pp (1981)

This is a revised version of 'The wired society', published in 1978. It deals with the impact of new technology and new communication systems on society. The final chapters on future developments, including the wired world, are of great interest.

166
Martin J and Butler D
'Viewdata and information society'
Prentice-Hall Inc 416 pp (1981)

The linking of telephone, television and computers to outside information sources is examined here. Part 1 is an introduction, Part 2 deals with technical developments and Part 3 surveys the impact of viewdata on specific industries (travel agents are one of the largest users). The second author is from Butler, Cox and Partners.

167
Moss M L
'Telecommunications and productivity'
Addison-Wesley Publishers Ltd 414 pp (1980)

This work examines the latest research on telecommunications systems with examples and case studies taken from private industry and public agencies. The emergence of new information systems is studied and their impact is described on both the office and the home.

168
Murphy B
'The world wired up. Unscrambling the new communications puzzle'
Comedia Publishing Group 155 pp (1983)

Individual chapters deal with the US, Canada, Europe, Japan, the UK and the Third World. This is an interesting attempt to see the concept of the wired nation in a global setting, and it is written in a non-technical language. Some countries are more widely covered than others; the section on Canada is both detailed and useful. This is the first attempt at a global view and as such is a useful introduction.

169
National Computing Centre
'Handbook of data communications'
National Computing Centre 497 pp (1982)

This work deals with the principles and concept of data communications and is an updated version of the Post Office Handbook.

170
Nilles J M, Carlson F R, Gray P and Hanneman G J
'The telecommunications-transportation tradeoff: options for tomorrow'
John Wiley and Sons Ltd 196 pp (1975)

This attempt to study the impact on new developments in communications on society, patterns of work, structure and location of business is a useful introduction to this complex but vital topic.
All the authors are from the University of Southern California.

171
OECD
'An exploration of the legal issues in information and communications technologies'
OECD Paris (Available from HMSO) (1984)

OECD have issued a series of reports on most aspects of telecommunications in Europe. This, the latest report, concentrates on the legal aspects and legal problems related to transborder data flow. The report includes a discussion of the OECD guidelines on agreed standards (accepted in 1980) on the protection of privacy and transborder exchange of personal data.

172
OEDC
'Telecommunications: pressures and policies of change'
OECD (Available from HMSO) 142 pp (1983)

This book reviews the implications of development of new technologies and services in relation to government policies in regard to competition at both national and international levels, investment, manpower and research and development.

173
Online Publications
'Business telecoms — the new regime'
Online Publications 197 pp (1983)

This work is based on the papers presented to an on-line conference in 1981 in London. It deals with integrated voice and data switching, new forms of electronic message transmission, local and wide area networks and the management techniques required for the new technology.

174
Online Publications
'Satellite communications'
Online Publications 142 pp (1980)

The impact of satellites on business communications is described here together with a review of recent development both in the US and Europe. It is based on the proceedings of an on-line conference held in London in 1980.

175
Online Publications
'Local network strategy and systems'
Online Publications 535 pp (1983)

This work includes 40 papers given at a conference entitled 'Localnet 83' which was held in London. It gives an overview of networks from both European and American viewpoints. Systems described include ETHERNET, Project Universe and Cambridge Ring networks. A section at the end deals with standards in the UK and Germany, as well as details of protocols.

176
Oyez Scientific and Technical Services Ltd
'Voice data integrated: what options confront the network designer?'
Oyez Scientific and Technical Services Ltd 160 pp (1983)

This volume is the published proceedings of a seminar held in London in December 1983. A wide-ranging view of current progress in the field of voice/data integration is given with emphasis on clarifying the options confronting the network designer. The papers are from a series of experts both from the UK and the US.

177
PACTEL
'Distributed database technology'
National Computing Centre 60 pp (1980)

A study commissioned by the Department of Industry in order to evaluate distributed database technology. Current European and US developments are examined and supplier policy discussed. PACTEL conducted this study in 1978.

178
Peebles P Z
'Communications systems principles'
Addison-Wesley Publishers Ltd 507 pp (1976)

A volume written for the graduate student which gives the basic theory of communication systems. There are many examples given of problems related to communications systems.

179
Pool I S
'Forecasting the telephone'
Ablex Publishing 192 pp (1982)

A continuation of the research on the telephone by a group led by the author at MIT. This book contains a review and analysis of 186 forecasts about the social impact of the telephone system from writings between 1876 and 1940. It forms a useful background for planning modern systems. There is an earlier work entitled 'The social history of the telephone' published by MIT Press.

180
Pool I S
'Technologies of freedom: one free speech in an electronic age'
Harvard University Press 299 pp (1983)

This work surveys the legal history of the differing modes of communications in the US and how free speech has been strengthened. As all of these separate functions such as telephones, television, radio and computers will be converging there is a danger of central control.

181
Roden M S
'Digital and data communications systems'
Prentice-Hall Inc (1982)

This is a comprehensive manual for the senior graduate. A knowledge of systems analysis is assumed, knowledge of communications systems is not. Many examples are given as an aid to learning.

182
Ross M H
'Fibre-optics networks or communication satellite: alternatives'
Arthur D Little Paris (Available from NTIS)
170 pp (1981)

Trends in communications technology are surveyed and the roles of fibre optics and satellite-based systems are explored. It is estimated that as costs fall fibre optics will become dominant. The main advantage of satellites is seen as a means of testing services before they are implemented in terrestrial networks.

183
Schiller D
'Telematics and government'
Ablex Publishing 237 pp (1982)

This work is divided into two parts: part 1 discusses in detail the history of the US Government telematic activities since the second world war and part 2 deals with the international scene. There is also an extensive bibliography. However, although there is a useful description of both the US and international scene no firm conclusions are given.

184
Scott P R D
'Modems in data communications'
National Computing Centre 166 pp (1980)

A description of the principles and practice of the use of modems is given and there are appendices which cover details of standards, connectors and CCITT/EIA terminology.

185
Scott P R D
'Reviewing your data transmission networks'
National Computing Centre 160 pp (1983)

This work gives a methodical approach to the assessment of the quality and cost effectiveness of the communications aspect of data networks. Performance and network management are included.

186
Shanmugam K S
'Digital and analogue communication systems'
John Wiley and Sons Ltd 624 pp (1979)

This work provides a detailed unified approach to both theoretical and practical aspects of digital and analogue communication systems. The approach is practical and no attempt is made to assess when digital systems will replace analogue systems.

187
Sereno K and Mortensen C D (editors)
'Foundations of communications theory'
Harper and Row Publishers Ltd 371 pp (1970)

This series of papers on research in communications theory covers the range of the technical, social and psychological aspects of communication, and the social network. One paper of interest links cybernetics with information theory. There is useful background information for the communications engineer: this advances the work of the classic papers by Shannon and Weaver.

188
Short J, Williams E and Christie B
'The social psychology of telecommunications'
John Wiley and Sons Ltd 206 pp (1976)

This is the first and only work to describe the socio-psychological aspect of interpersonal telecommunications. Both the beneficial and harmful effects that could arise with the increased use of communications systems are dealt with. The effect of visual communications and the possible development of telecommuting are described and conclusions for future implications are given.

189
Singh I B
'Telecommunications in the year 2000'
Ablex Publishing 221 pp (1983)

This work consists of 25 papers from a telecommunications conference at Rutgers University in November 1981. Arranged in five sections they cover global outlook, policy, information systems, rural development and prospects.

190
Swan A J
'Data communications protocols'
National Computing Centre 155 pp (1978)

This work covers equipment interconnections, character structure and data link control protocols, packet-switching networks, the CCITT X.25 interface standard, communications software, network architecture and user considerations. This work is intended for both system designers and users.

191
Schwartz M
'Computer-communications network design and analysis'
Prentice-Hall Inc 372 pp (1977)

The design of data communication systems is described and practical examples are given such as airline reservations and banking networks, private corporate networks and commercial timesharing.

192
Tanenbaum A S
'Computer networks'
Prentice-Hall Inc 517 pp (1981)

This introduction to computer networks places its emphasis on network architecture and protocols from the Physical Layer to the Application Layer and from local area networks to satellite networks. ARPANET, SNA, DECNET and X.25 are used as examples throughout the book.

193
Veljanoviski C G and Bishop W D
'Choice by cable — the economics of a new era in television'
(Hobart paper 96)
Institute of Economic Affairs 115 pp (1983)

The demand for global satellite systems such as INTELSAT and ARABSAT is predicted both for the countries with a well developed aerospace industry as well as their world-wide use. An estimate is made that by the year 2000 there will be up to 475 satellites and five platforms in use.

194
Watts R A
'Interactive computing for the lay user'
National Computing Centre 150 pp (1983)

An introduction to the design of interactive computer systems covering the choice of dialogue style, user/computer interface and user involvement.

195
Weck M H
'Communications Standard Dictionary'
Van Nostrand Reinhold 1045 pp (1982)

An extensive compilation of terms used by designers, developers, manufacturers, vendors, users, managers, administrators, operators and maintainers of communication systems and components.

196
Williams F
'The communications revolution'
Sage Publications 292 pp (1982)

The author, founder of the Annenberg School of Communications at the University of Southern California, describes current development in communications and how future development will affect society in general and human behaviour specifically.

197
Wilson P
'The electronic mailbox'
National Computing Centre 200 pp (1983)

Electronic mailbox systems will start to rival telephone systems in the 1990s and beyond. This book describes what they are, how they can be designed and what they can be used for.

198
Winsbury R (editor)
'Viewdata in action: a comparative study of Prestel'
McGraw-Hill Book Co (UK) Ltd 237 pp (1981)

A series of articles under subject headings, each with an introduction by the editor. The relationship of viewdata to technology and industry, electronic publishing and databases is covered. The concluding section is a survey of similar development in the rest of the world.

199
Woolfe R
'Videotex — the new television/telephone information services'
Heyden and Son Ltd 170 pp (1980)

This work gives an overview of videotex and the current situation and an assessment of the market. There is a section on Prestel, development in Europe, US and Japan.

Keeping up to date

Journals of interest

'Communicate for the telecommunications user'
Communicate Publications Ltd

Monthly. Covers current developments.

'Networks'
John Wiley and Sons Inc

An international journal published quarterly. Subject coverage is in the areas of computer networks, telecommunications systems, traffic networks and electronic networks.

'Telecommunications policy'
Butterworth & Co (Publishers) Ltd

International quarterly journal. Covers all topics dealing with the development of the new technology such as video conferencing and networking. It also has a good book review section.

'Telematics and informatics'
Pergamon Press

An international journal covering research and review articles on new developments in the relationship between computers and information. Published quarterly.

Reference sources

In the list of bibliographic references above journal titles are abbreviated as little as possible. The full title is given below together with the address from which the publication is available. Book publishers and other sources are also included in the list.

Ablex Publishing Corporation
355 Chestnut Avenue
Norwood
NJ 07648
US

(UK Distributors:
International Book Distributors
56 Wood Lane End
Hemel Hempstead
Herts
UK)

Academic Press Inc (London) Ltd
24 Oval Road
London
NW1 7DX
UK

Addison-Wesley Publishers Ltd
Finchampstead Road
Wokingham
Berkshire RG11 2NZ
UK

American Management Association
135 West Fiftieth Street
New York
NY 10020
US

Architectural Press Ltd
9 Queen Anne's Gate
London SW1H 9BY
UK

British Library
Publications Section — Lending Division
Boston Spa
Wetherby
West Yorkshire LS23 7BQ
UK

Butterworths and Co (Publishers) Ltd
Subscriptions Dept
Oakfield House
Perrymount Road
Haywards Heath
Sussex RH16 3DH
UK

Comedia Publishing Group
9 Poland Street
London W1V 3DG
UK

Communications and Information Technology (CIT) Research Ltd
Circus House
21 Great Titchfield Street
London W1P 7FD
UK

Communicate Publications Ltd
Link House
Dingwall Avenue
Croydon CR9 2TA
UK

Department of the Environment and Transport
2 Marsham Street
London SW1P 3EB
UK

Economic Intelligence Unit
Spencer House
27 St James Place
London SW1A 1NT
UK

Ekistics
Athens Centre of Ekistics
PO Box 3471
IQZIO Athens
Greece

Ellis Horwood Ltd
(Subsidiary of John Wiley and Sons Ltd)
Baffins Lane
Chichester
West Sussex
UK

Euromonitor Publications
18 Doughty Street
London WC1N 2PN
UK

Future Systems Inc
16726 Oakmont Avenue
Gaithersberg MD
US

The Greater London Council
County Hall
London SE1 7PB
UK

HMSO
Main Bookshop
49 High Holborn
London WC1V 6HB
UK

(Mail order:
PO Box 276
London SW8
UK)

Harper and Row Publishers Ltd
28 Tavistock Street
London WC2E 7PN
UK

Harvard University Press
The University Presses of Chicago, Harvard and MIT
126 Buckingham Palace Road
London SW1W 9SD
UK

Heyden and Son Ltd
Hillview Gardens
London NW4 2JQ
UK

Institute of Economic Affairs
2 Lord North Street
London SW1 3LB
UK

Kegan Paul International
(Subsidiary of Routledge and Kegan Paul Ltd)
Broadway House
Newton Road
Henley-on-Thames
RG9 1EN
UK

Learned Information
Besselsleigh Road
Abingdon
Oxford
OX13 6LG
UK

Lexington Books
125 Spring Street
Lexington
MA 02173
US

(UK Distributors:
Gower Publishing Co Ltd
Gower House
Croft Road
Aldershot
Hants
UK)

Logica Ltd
64 Newman Street
London
W1A 4SE
UK

McGraw-Hill Book Co (UK) Ltd
Shoppenhangers Road
Maidenhead
Berkshire
SL6 2QL
UK

Macmillan Press
Brunel Road
Houndmills
Basingstoke
Hants
RT21 2XS
UK

The MIT Press
126 Buckingham Palace Road
London SW1W 9SD
UK

van Nostrand Reinhold Co Ltd
Molly Millars Lane
Wokingham
Berks
UK

NTIS United Kingdom Service Centre
PO Box 3
Alton
Hampshire
GU34 2PG
UK

National Computing Centre
Oxford Road
Manchester
M1 7ED
UK

North-Holland Publishing Co
PO Box 211
1000 A E Amsterdam
The Netherlands

OECD (Organisation for Economic Cooperation and Development)
2 rue Anxre-Pascal
75775 Paris Cedex 16
France

Online Publications Ltd
Pinner Green House
Ash Hill Drive
Pinner
Middlesex
HA5 2AE
UK

Oyez Scientific and Technical Services Ltd
Bath House
56 Holborn Viaduct
London
EC1
UK

Penguin Books Ltd
Bath Road
Harmondsworth
Middlesex
UB7 0DA
UK

Peter Peregrinus, Publishers
PO Box 26
Hitchin
Herts
SG5 1RJ
UK

Pergamon Infotech Ltd
Berkshire House
Queen Street
Maidenhead
Berks
SL6 1NF
UK

Pergamon Press
Headington Hill Hall
Oxford
OX3 0BW
UK

Prentice-Hall Inc
66 Wood Lane End
Hemel Hempstead
Hertfordshire
HP2 4RG
UK

Quiller Press Ltd
11a Albemarle Street
London
W1X 3HE
UK

St Martin's Press Inc
175 Fifth Avenue
New York
NY 10010
US

(UK Distributors:
Columbus Books
29 Elmfield Road
Bromley
Kent
UK)

Sage Publications Ltd
28 Banner Street
London
EC1Y 8QE
UK

and

275 South Beverley Drive
Beverley Hills
CA 90212
US

Charles Scribner's Sons
597 Fifth Avenue
New York
NY 10017
US

Charles Scribner's Sons European Distributors:
European Book Services
Flivdaan
36-38 PO Box 124
1380 AC Weesp
The Netherlands

Telecommunications Press
(Subsidiary of Architectural Press Ltd)
9 Queen Anne's Gate
London
SW1H 9BY
UK

John Wiley & Sons Ltd
(Book orders)
Baffins Lane
Chichester
Sussex
PO19 1VD
UK

John Wiley & Sons Ltd
(Periodical orders)
605 Third Avenue
New York
NY 10016
US

World Future Society
4916 St Elmo Avenue
Betjesda
MD 20814-5089
US

Index

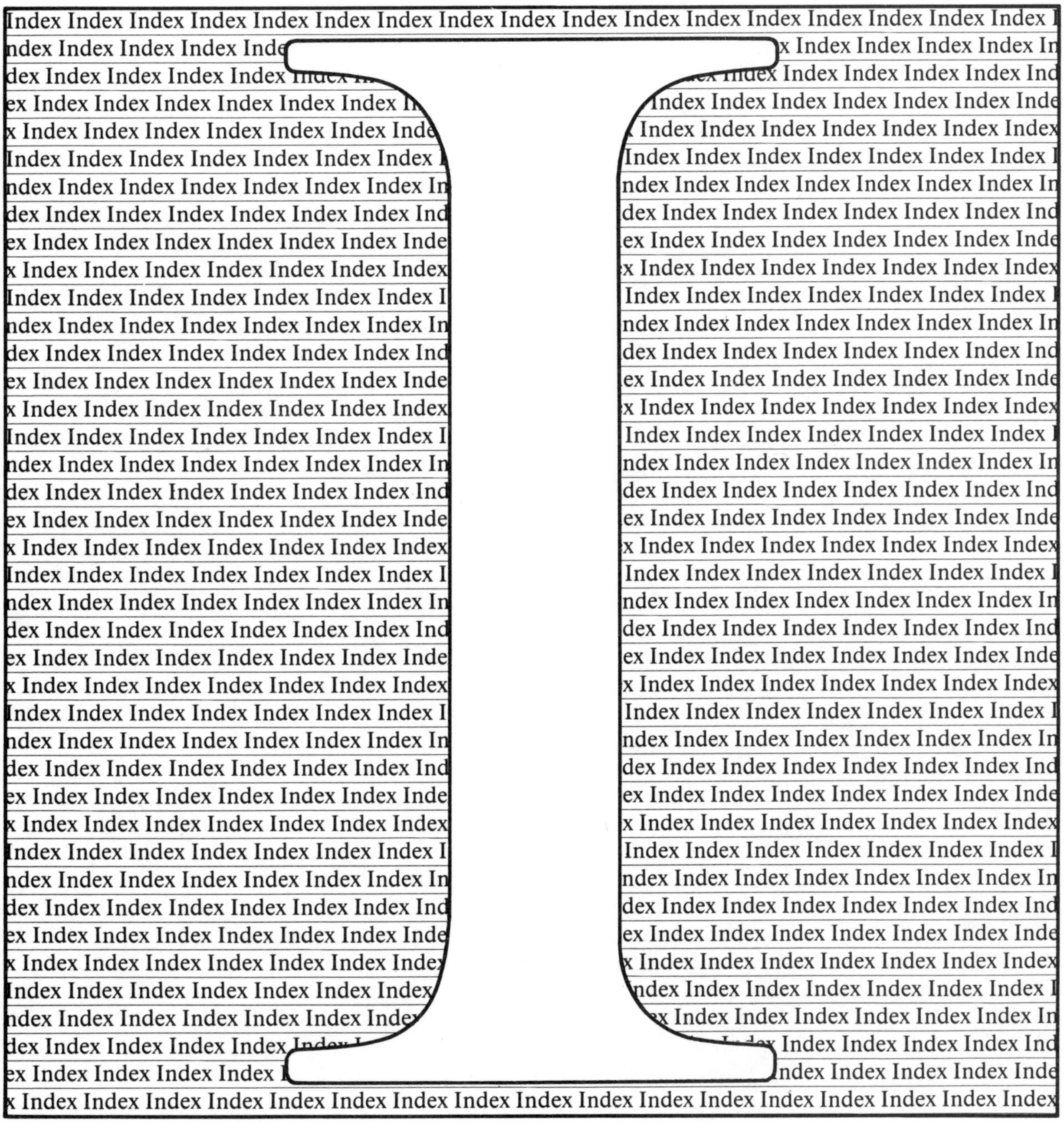

Subject and contributor index